AF711445

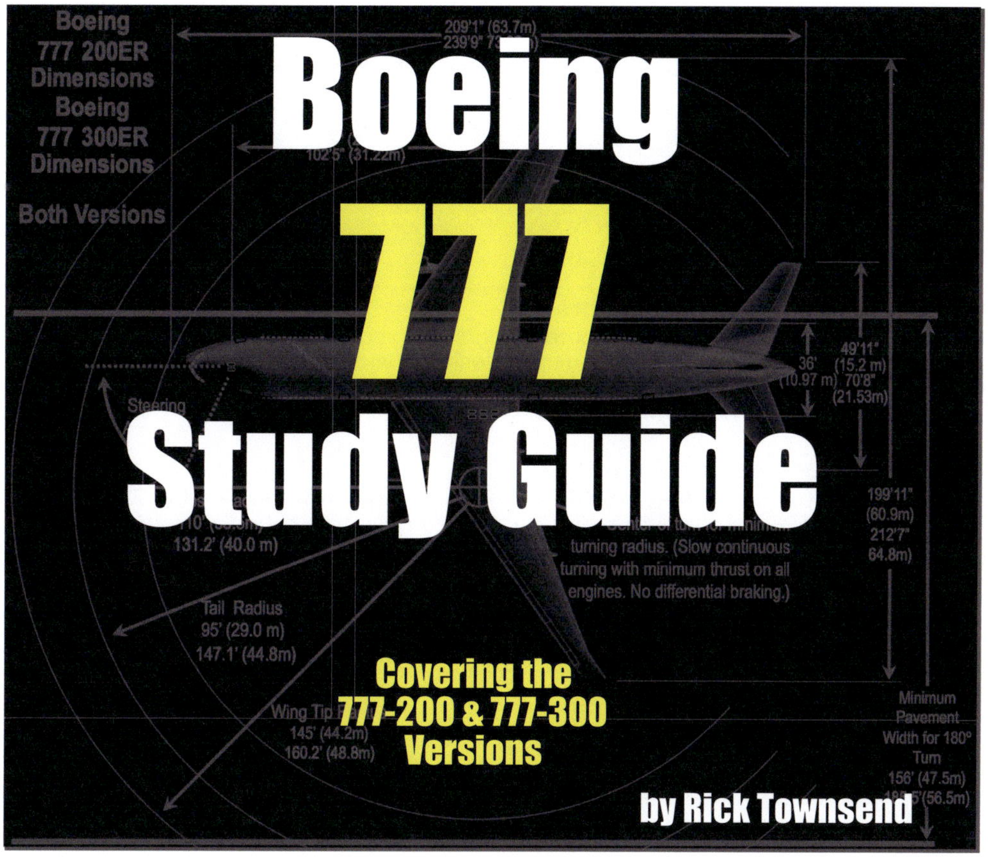

This book was originally developed for pilots at American Airlines. Neither American Airlines nor Boeing Corporation have endorsed or in any way been responsible for the content of this book.

As a courtesy to the pilots of American Airlines who have supported the development of these books, it is noted that this edition is current to the best ability of the author to make it so, and includes material relevant to:

Flight Manual Revisions Incorporated:

Manual	Revision	Abbreviation
Aircraft Operating Manual	2.2*	AOM
Aircraft Systems Manual	05	ASM
Quick Reference Handbook	44	QRH
Flight Operations Manual	20*	FOM

* **Note**: *After FOM Revision 20 and AOM revision 02, FOM and AOM are password-protected and cannot be updated for review by the author.*

Revision Notes

January 1, 2023 Edition—(Initial issue) Most recent updates in this edition:

OM1, OM2 and FM1 data has migrated to AOM, ASM, and FOM respectively. Page number references in this Study Guide have been validated and moved to the new page references.

Page 11 – Added information related to fuel system AD 2019-16-13; reformatted fuel limitations section to accommodate new data.

March 24, 2023 Edition—(Initial issue) Most recent updates in this edition:

Page 19 – Updated page reference for AC to Pack takeoff from Performance Manual to appropriate AOM page.

June 5, 2023 Edition—Updates in this edition:

Page 29 – NO AUTOLAND and LAND2 notes updated.

Page 37 – ENG THRUST (L or R) logic description updated.

August 14, 2023 Edition—Updates in this edition:

Page 6 – Limitations Section—Numbering added for those readers whose carriers use section 18 numbering convention.

Page 11 – Added notes indicating removal of restrictions associated with ADs 2023-10-02, 2023-12-13 and 2023-12-11 related to radio altimetry restrictions and interference.

Page 11 – Added Holding Speeds for ICAO as well as FAA airspace.

November 2, 2023 Edition—Updates in this edition:

Page i – Note added indicating to readers that after FOM Revision 20 and AOM revision 02, FOM and AOM are password-protected and cannot be updated for review by the author. *Use this guide at your discretion. Manufacturer or Air Carrier guidance always takes precedence.*

Published by Pilot Study Guides, LLC, Grapevine, Texas.
First Edition and Copyright 2004
Updated continuously (including copyright protection) through the current 2023 edition.

Graphics: With one exception, all graphics were produced by and are the copyright-protected property of the author, except where noted within this edition.

Acknowledgement: Thanks to Norebbo Design Studios, *www.norebbo.com* for allowing use of the profile drawing in the *Airplane General* section.

All rights reserved. No part of this publication may be reproduced, distributed, or transmitted in any form or by any means, including photocopying, recording, digitally reproducing, or using any other electronic or mechanical records, without the prior written permission of the copyright holder. Exception is granted for reproduction of checklist portions of the book where so dedicated on specific pages.

Corporation References: American Airlines™ and the Boeing Corporation™, referenced herein, have neither endorsed nor provided any proprietary information for use in these books. Their referencing is for informational awareness only.

ISBN: **978-1-946544-46-9** (Print)
ISBN: **978-1-946544-48-3** (Amazon Kindle)

Table of Contents

Introduction

What This Book IS

This Study Guide is a compilation of notes taken primarily from the flight manual, but also includes elements taken from class notes, computer-based training, and operational experience. It is intended for use by initial qualification crewmembers preparing for orals, and also for systems review prior to recurrent training or check rides. It is assembled in an attempt to organize in one location all the buzz words, acronyms, and numbers the average guy needs to know in order to get through the events above from an aircraft systems standpoint.

What this Book IS NOT

This series of books was originally developed for use by American Airlines pilots. It is for that purpose that operating manual references are for the manual sets for that airline as a courtesy to my loyal customers of many years. The books ***are** <u>**not**</u>* officially sanctioned in any way by any airline, however. The author assumes all responsibility for accuracy. (Forward corrections to the address provided, please!)

These books ***do not*** replace study of the operations manual, but instead provide a supplementary source of review material to complement study of official publications. Except for a few specific areas, it does not include regulatory guidance or other materials—just airplane systems details, some ideas for organizing the cockpit flow, cold weather operations, etc. The Study Guide is not printed on fancy paper or expensively bound. That makes it easy to tuck into your kit bag or briefcase and study as you travel. Therefore, for these same reasons, it is far less expensive than most commercially produced books of its kind. It is intended to be affordable and usable.

Suggested Uses

A good time to use this book is on layovers. A reasonable plan would be to attempt to review limitations, memorization Red Box items once per sequence, and the rest of the sections once per month. Reviewing sectional study outlines usually leads to trying to remember one of those great plumbing diagrams from the flight manual or synoptics, and so those also get reviewed in the process.

Limitations and Black Border items are laid out so that you may cover the answer with a 3x5 card and quiz yourself. Other sections are in outline form.

A word about acronyms—you will probably see most of the acronyms you have heard of before in this booklet, as well as a few new ones. Most of us avoid them when possible, using clues on aircraft instrument and systems management panels to jog our memories. But for those you have trouble with, I've included all the ones I could think of. It is not necessary to learn all of them, but only those you need to help remember tough sets of items. Thankfully, the 777 fleet management folks believe in keeping these "laundry lists" to a minimum, so you will see fewer of these than you will likely have seen in previous aircraft types.

Unique Formatting Features

Throughout this book, several specific methods are used to highlight information. They include:

Recommended memorization sections

A key feature in the study guide includes highlighting information which is recommended for memorization. These are areas that have been asked on orals or which simply need to be known without reference to the books. This new feature is found in systems notes, but not in the limitations section, which all requires a high level of retention. If you have recommendations for information that should be so highlighted but is not, please let me know and I will do so in a future version.

Master Caution and Master Warning lights are Indicated by highlighted MC (Master Caution) and MW (Master Warning) highlights within this Guide.

Panel or Warning Light Text—Words or indications found on system panels or switches is indicated by bold, all capital letters such as **LOW PRESSURE**.

300ER Series data has been included throughout this book. Editorial decisions have been made about how much 300ER data to include. Your inputs are gratefully accepted.

Aircraft versions are indicated in this guide where differences occur with the notation 200ER or 300ER. Where not specified, data pertains to both aircraft models. Only where it is different will the identifying tags be used. Thanks for your assistance in letting me know if you feel additional data on either model should be included or corrected.

Electronic Version

An electronic version of this study material is also available. Buttons at the top of each page represent functions available in the iBooks version of this study guide (iPad/smart phone) to return to the Table of Contents or panel drawings. See web site for details.

Table of Contents | Overhead Panel | Fwd Panel Ctr Console

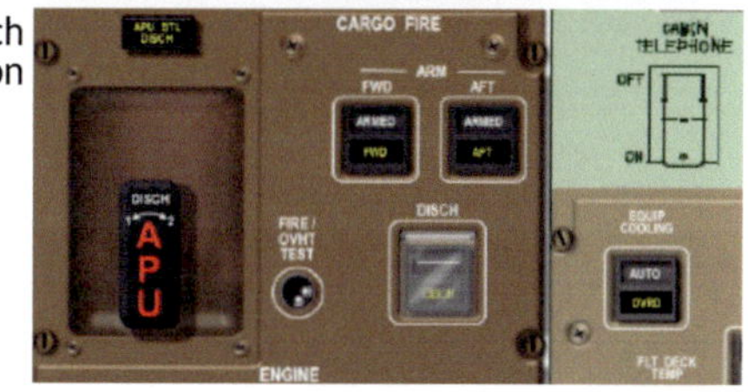

The corner boxes on each page identify the book section and are not hyperlinked.

Panel drawing pages show graphics which are described in this book. Clicking on or touching the colored graphic panels in electronic versions of this book moves to the page with that graphic and its description.

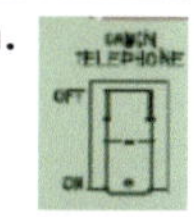

Green shaded areas are not depicted or discussed in this book.

Reproduction

Please don't photocopy this book. I have made a serious effort to keep the costs low for you. In fact, you can buy this book for about half the price of comparable material offered through other commercial publishers. There is a lot of work in this book, as you'll soon be able to see when you study the material. Pages with notes, when provided and annotated as such, may be freely copied by customers. Thanks for your cooperation.

Again, please send me any suggestions or corrections using the contact information below:

Rick Townsend ....Grapevine, Texas
....(817) 416-5533
E-Mail.. rick@pilotstudyguide.com
Web Site.. www.PilotStudyGuide.com

Revisions & Subscriptions

Revisions are made on a regular basis. The downside of revisions is that updating your digital version causes you to lose highlighting and notes you have made. This is because each version includes a change to the printing date, so each page is updated on every new release. The methods for updating digital versions varies by platform. See the Pilot Study Guide web site for details.

To contact the Flight Academy Gift Shop and have a book mailed to your home, you may use the following:

Office 817-967-5540
Via Internet.. www.crewoutfitters.com

Publication Details

Compiled: January 23, 2025

All rights reserved.

Published by
Pilot Study Guides, LLC, Grapevine, Texas

© 2004 by Rick Townsend, Updates 2005-2023
Cover Graphic © 2018 by Rick Townsend

For ISBN and additional publication details, see title page i

Panels

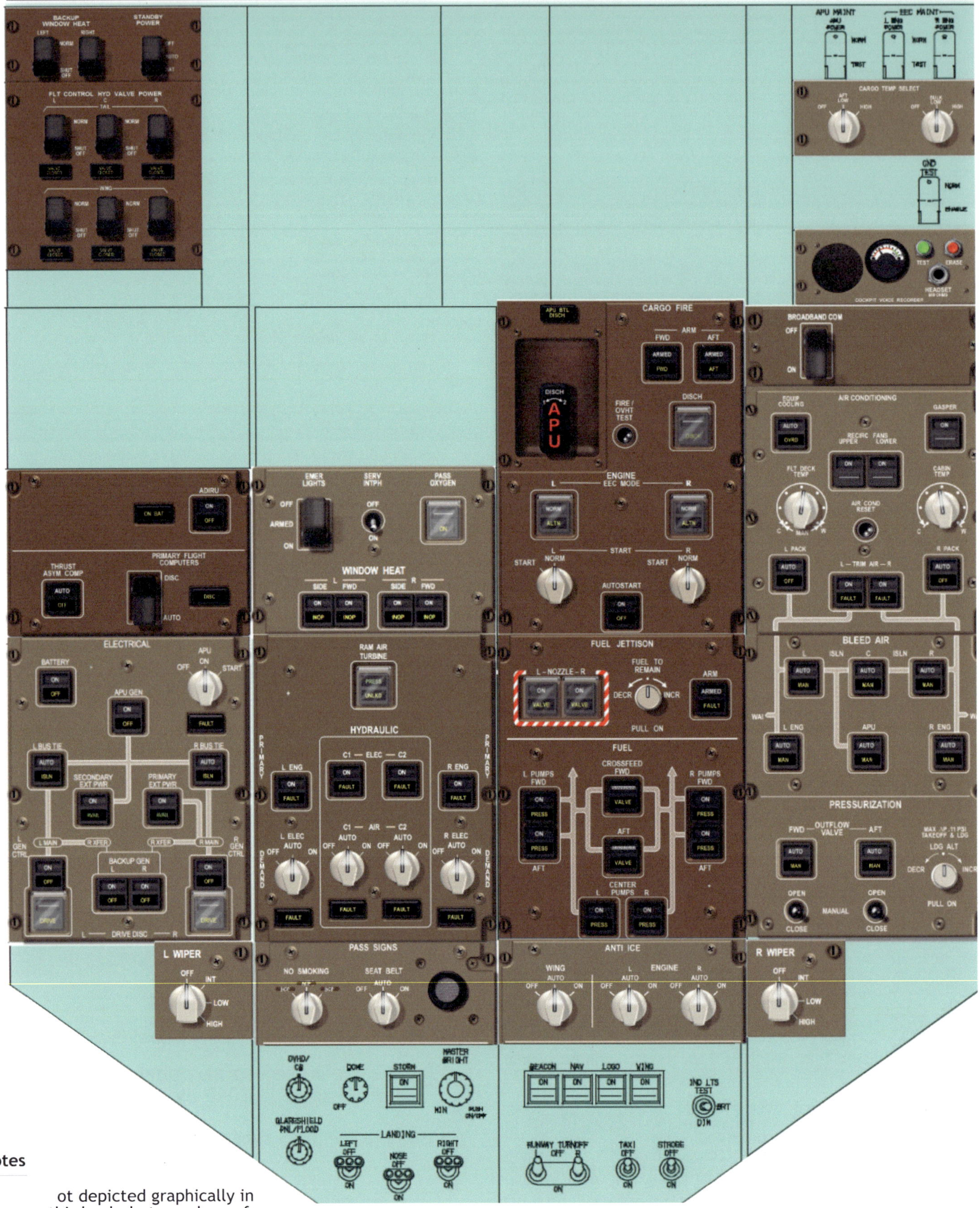

Notes

ot depicted graphically in this book, but are shown for orientation.

♦ Electronic version: Touch the colored panel to jump to the page with that system and panel description.

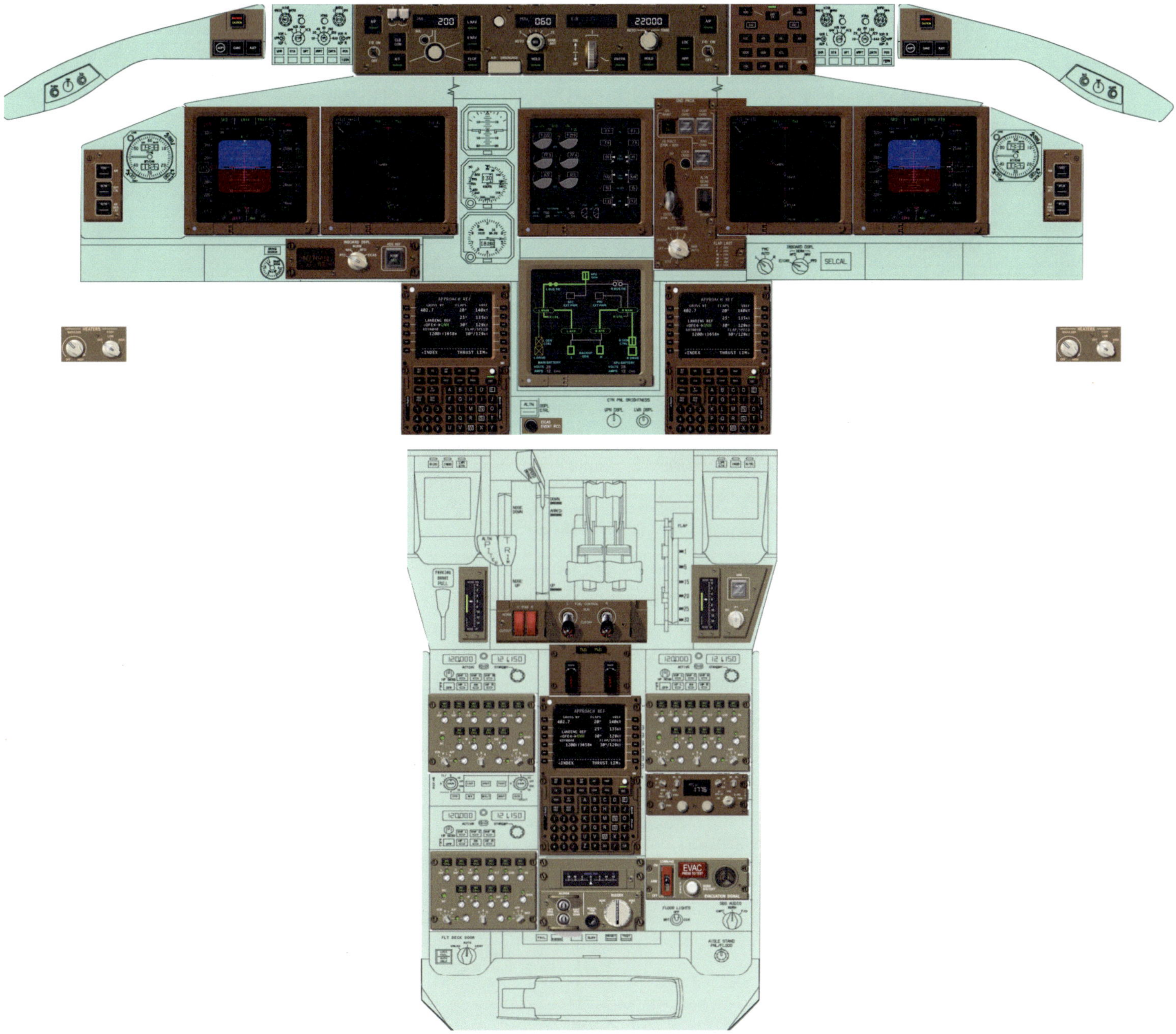

Notes:

- ♦ Areas in tinted green are not depicted graphically in this book, but are shown for orientation.
- ♦ Electronic version: Touch the colored panel to jump to the page with that system and panel description.

Limitations

This Section is laid out for study with system areas on one side and the appropriate limitation on the other, so you can cover the answer with an index card and quiz yourself. Some of the limitations in Section 2 are not often encountered, such as alternate fuel limitations. For that reason, they have been shortened here for study. These will be listed in the left column, so that as you study, you will be reminded the limitation exists in the book and will (hopefully) think to look there should the situation be encountered. The Limitation column contains only a reminder to check the Ops Manual for the item in question.

Memorization items are in ***bold italic prin***t for ease of reference.

Crewmembers are required to be familiar with all items covered by a limitation. *fleet does not have memorization dashed boxes in the Limitations section, however.*

Items formatted with green background are limiting guidance from other sections of the operating manual or Flight Operations Manual (FOM). They are included in this section for ease of study.

All items pertain to all models unless indicated to be different with the model tags 200ER and 300ER.

▲ Per AOM convention, all **AFM** items are indicated with a gray triangle symbol

18.1 General Limitations

Certification Status▲: 777 aircraft are certified Transport Category, FAR Part 25 & FAR Part 36.▲

Airplane Configuration▲: Aircraft configuration must not vary from those described in the AOM or the QRH non-normal procedures.
Also, refer to Flightdeck Forms Book (Yellow Book).
The aircraft configuration specified in the AOM for various flight regimes is the configuration on which the allowable weights given in the performance sections are based. Therefore, the prescribed configurations must be used to assure the aircraft performance requirements will be met.

Types of Aircraft Operation▲: Airplane approved by the FAA for both day and night in visual, instrument, icing, and extended over water operations.

18.2 Operational Limitations

Altitude Limits ▲

Maximum Certification Altitude, Flight: 43,100 Feet
T/O & Landing: 8,400 Feet
Maximum Operating Altitude with flaps and/or slats extended: 20,000 feet
RVSM Altimeter Tolerances CA — FO Displays: 200' maximum allowable for RVSM operations
Before T/O, max difference, CA or FO Display vs. Field Elevation: ±75 feet
STBY Altimeter Use for RVSM: STBY Altimeter does not meet RVSM accuracy requirements.

Speed Limits ▲

		200ER	300ER
Maximum Operating Airspeeds	**V_{MO}:**	330 Knots IAS	330-350 Knots IAS (Varies by altitude, see chart below
Maximum Operating Mach	**M_{MO}:**	0.87 Mach	0.89 Mach

Determined By: Lower edge of red and black colored region of the speed tape on the Primary Flight Displays (PFD).
Limits shall not be deliberately exceeded in any regime of flight

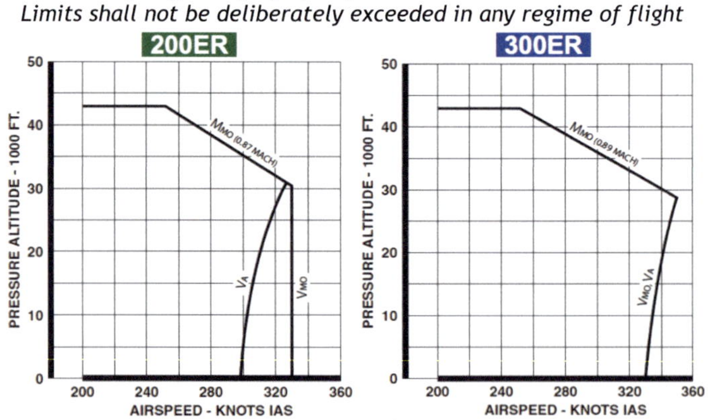

Severe Turbulence Air Penetration Airspeed
Below 25,000 feet: ***270 Knots***
At or Above 25,000 feet: ***280 knots / .82 Mach [1]***

[1] Maintain a minimum speed of 15 knots above the minimum maneuvering speed (amber band) at all altitudes when the airspeed is below 0.82 Mach.

Maximum Gear Operating Speed V_{LO}: 270 knots / .82 Mach
Maximum Gear Extended Speed V_{LE}: 270 knots / .82 Mach

Design Maneuvering Speeds V_A Definition the speed above which maneuvers involving full application of rudder, ailerons or elevator, or maneuvers involving angles of attack near stall, should be avoided.

Maximum Flap Operating Speeds

	200ER	300ER
1:	255	265
5:	235	245
15:	215	230
20:	195	225
25:	185	200
30:	170	180

Weight Limits ▲

Notes: Maximum takeoff and landing weights may be further limited by runway analysis. See performance sections for maximum permissible operating weights for specific runways, weather conditions and abnormal aircraft configurations. Takeoff weight may be further restricted by the minimum equipment list and/or configuration deviation list. Maximum weight limits also may be further limited by center of gravity, fuel density and fuel loading limitations.

Note: Alternate maximum weights are placarded on the flight deck.

Maximum Takeoff Weight: The Maximum Takeoff Weight for a given flight is the lightest of the:
- ✦ Climb Limit
- ✦ Brake Energy Limit
- ✦ Field Length Limit
- ✦ Tire Speed Limit
- ✦ Obstacle Limit
- ✦ Structural Takeoff Weight Limit

All of these limits, except for the structural limit, are defined by the performance capability of the aircraft. See OM 1p.7.2

Structural Limit: Weight limit based on the weight capability of the aircraft structure.

Overweight Landing Definition: An overweight landing is defined as a landing made at a gross weight in excess of the aircraft's maximum structural landing weight.
An overweight landing above the maximum certified landing weight is considered an emergency situation and may compromise landing safety margins, however, it is considered a safe practice when complying with overweight landing procedures.

Maximum Landing Weight: The lightest (most limiting) of the following maximum landing weights:
- ✦ Structural Landing Weight Limit
- ✦ Landing Climb Limit
- ✦ Landing Field Length (Runway) Limit
- ✦ Approach Climb Limit

In the Landing Section, only one climb limited weight is shown. It is the most restrictive (lightest) of the approach climb or landing climb limit.
Limiting weights are described on FOM 1p.7.10

FOM References:
- ✦ FOM 11n.2 Overweight Landing for the overweight landing policy
- ✦ FOM 11n.2.2 for AML Writeup requirements.
- ✦ QRH Miscellaneous 0.50 for the Overweight Landing Checklist

Weight Limits ▲	200ER	300ER
Maximum Taxi Weight:	***650,000 lbs.***	***777,000 lbs.***
Maximum Takeoff Weight:	***648,000 lbs.***	***775,000 lbs.***
Maximum Landing Weight:	***460,000 lbs.***	***554,000 lbs.***
Maximum Zero Fuel Weight:	430,000 lbs.	524,000 lbs.
Minimum In-flight Weight:	266,100 lbs.*	305,500 lbs.* *Does not include usable fuel

Note: Refer to FOM> Landing> Non-Normal/Emergency> Overweight Landing. Alternate maximum weights are placarded on the flight deck.

Wind Limitations (Knots)	200ER	300ER
Maximum Wind for Takeoff and Landing	50	
Max Tailwind Component for T/O & Landing	***10*** [1]	***15*** [1]
Autoland—Headwind	25	
Autoland—Tailwind	10	15
Demonstrated Crosswind	***38***	
Takeoff Crosswind Limits		
Dry	38	
Wet		
Standing Water/Slush	25	
Snow (Wet & Dry)		
Compacted Snow	20	
Ice	15	
RCC—Pirep Landing Crosswind Limits		
6—Dry	38 [2]	
5—Good		
4—Good to Medium	35 [2]	
3—Medium	25 [2]	
2—Medium to Poor	17 [2]	
1—Poor	15 [2]	
Autoland	25	
Visibility < 3/4 SM (RVR 4000) (1200 Meters)	15	

Notes

1. May be further limited by runway or visibility limitations.
2. Reduce crosswind limits by 5 knots on wet or contaminated runways whenever asymmetric reverse thrust is used. See 5 Knot Cross Wind Limit Reduction.

The 5 knot reduction is intended to be used with the landing RCAM/RCC codes or ECL/QRH crosswind limitations.

The 5 knot reduction does not need to be applied to crosswind limits that are reduced only for visibility less than (3/4 sm [RVR 4000] [1200 meters]) or for autoland limitations.

In addition, it only applies if greater than idle reverse thrust is used on the engine with the operating reverser.

Performance permitting, if idle reverse thrust can be maintained on the engine with the operating reverser the 5 knot reduction can be avoided.

Runway less than standard width (148 feet / 45 meters): 20 knots FOM 5g.1.4

Ground Wind Limits

Ground Wind Operating Envelope (300ER)

Tailwinds & crosswinds between 30-45 Knots: Engine power should be limited to 70% N1 maximum.

Tailwinds & crosswinds over 45 Knots: Avoid thrust above that required for normal taxi operation.

Item	Condition	Limit
All Doors	Opening or Closing:	40
	While Open:	65
Miscellaneous Limits ▲	Runway Slope:	±2%
	Minimum Flight Crew:	Captain, First Officer
Autobrake Requirements	When Required:	♦ Runway <7000' ♦ RVR <4000' or visibility <3/4 mile (1200m) ♦ Runway contaminated with standing water, snow, slush, or ice ♦ Approach speed increased above normal by procedure. ♦ Braking conditions less than good AOM 9.1.1, p. 9-3, Arrival Tasks
	When Recommended:	Landing with gusty winds or crosswinds AOM, 12.2.1, p. 12-28; 15.7.7

18.3 Aircraft General

Item	Condition	Limit
Flight Crew Rest Facility ▲	300ER:	Flight Crew Rest (OFCR) Area and Overhead Flight Attendant Rest (OFAR) Area are to be occupied by crew members only if they have been trained in use of the crew rest evacuation routes, firefighting procedures and depressurization procedures.
Bottle Pressure	Reference Temperatures:	For limitations associated with bottle pressures, use ambient temperature as representative of bottle temperature.
Passenger Evacuation ▲	Prior to Taxi, T/O & Landing with passengers:	♦ ***Main door emergency power assist systems and emergency evacuation slide systems must be armed with Mode Select Handles in the ARMED position.*** ♦ ***Girt bar indication flags visible through windows in doorway liners when the doors and slides are in the ARMED mode.***
	Cabin Phone Handsets	Must be stowed, and the cabin prepared to satisfy direct view requirements by lowering passenger seat headrests and opening the partition viewing window, prior to taxi, T/O and landing whenever passengers are carried.
Flightdeck Door & Access System ▲	Checks:	System check must be accomplished prior to First Crew Flight of the day.
Weather Radar	Inside Hangars:	Avoid weather radar operation in a hangar or within 50 feet of fueling operations or fuel **NOTE:** Hangar recommendation does not apply to radar test mode.
	Avoid Weather Radar Operation When ▲:	Personnel within area normally enclosed by aircraft nose radome.
Configuration Deviation List (CDL) ▲	When operation is scheduled with certain secondary airframe & engine parts missing:	The airplane must be operated in accordance with the CDL certificate limitations contained in the CDL section of the Minimum Equipment List Manual
Oxygen Bottle Minimum Pressure	ETOPS:	1250 psi at 70°F
	Non-ETOPS:	800 psi at 70°F AOM, 3f.2.3, p. 3.f-8, Pre-Flight

18.4 Air Systems

Item	Condition	Limit
Cabin Pressurization ▲	Maximum Cabin Pressure Differential for Takeoff and Landing:	0.11 psi
	Maximum Cabin Pressure (Relief Valve):	9.1 psi
APU to Pack Operation ▲	300ER:	APU to Pack takeoffs prohibited at airport altitudes > 6900' field elevation

18.5 Anti-Ice, Rain

Item	Condition	Limit
Icing Conditions ▲	Definition:	♦ ***OAT 10°C (50°F)*** and ♦ Visible moisture in any form is present (clouds, fog with visibility one mile or less, rain, snow, sleet, or ice crystals.)
	In Addition, During Ground Ops and Takeoff:	When operating on ramps, taxiways, or runways where surface snow, ice, standing water, or slush may be ingested by engines, or freeze on the engines, nacelles, or engine sensor probes.
	And in Addition, In Flight:	***When indicated by the primary ice detection system or 10°C or below TAT.***
Anti-Ice Systems ▲ Engine Anti-Ice	Ground:	Must be ON if icing conditions exist or are anticipated **CAUTION**: Do not use Engine anti-ice when OAT is above 10°C
	Flight:	Must be ON or AUTO if icing conditions exist or are anticipated
	Exception In Flight:	OAT Temperature is below -40°C (SAT).
	Do Not Use Anti Ice When:	**CAUTION**: Do not use Engine anti-ice when TAT is above 10°C **WARNING**: • Do not rely on airframe visual cues to turn engine anti-ice on. • Use temperature and visual moisture criteria as defined above. • Delaying use of engine anti-icing until ice buildup is visible from cockpit may result in severe engine damage and/or flameout.
Wing Anti-Ice	Flight:	**CAUTION**: Do not use wing anti-ice when TAT is above 10°C

18.6 Autoflight

Radio Altimeter ▲ **Flight Restrictions:**	See AOM 18.13.2
Approaches and Landings in presence of Radio Altimeter 5G C-Band Interference:	See AOM 18.13.3
FLCH **Final Approach Min Engagement Altitude:**	1000 feet AFE
Aileron Trim **Use with Autopilot Engaged:**	Prohibited
Autopilot / Flight Director System	
Minimum Altitudes for Autopilot Use ▲ **Takeoff:**	200 Feet AFL
Coupled to ILS Glideslope and localizer	
LAND 2 or LAND 3 Annunciated:	None
Without LAND 2 or LAND 3 Annunciated:	200 Feet AFL
Go Around Mode:	None
Automatic landings	
Flap Settings ***May be made with flaps:***	***20 or 30***
Category II/III Autolandings with Flaps 25 ▲**:**	Not approved
Engines ***Single or dual engine Operation:***	***Both operating OR one engine inoperative***
Annunciator **Autopilot Flight Director System (AFDS) Autoland** ***Status Annunciation:***	***Must display*** LAND 2 or LAND 3
May NOT be Made with ***EICAS:***	***These messages, if displayed, prohibit automatic landings:*** ♦ SLATS DRIVE ♦ PITCH UP AUTHORITY
Glideslope Angle ▲ **Maximum angle:**	3.25°
Minimum angle:	2.5°
Wind:	See Operational Limitations, *page 6* of this book or AOM 18.2.4.

18.7 Auxiliary Power Unit (APU)

APU **Starter Duty Cycle (Not AFM)**

Maximum Start Attempts Within 60 Minutes:	Three attempts maximum	
	Electric Starter Motor Wait:	**For Turbine Starter Wait:**
Between 1 & 2:	1 Minute	1 Minute
Between 2 & 3:	1 Minute	1 Minute

APU Start/Shutdown During Refueling / Refueling: WARNING Do not start or stop the APU should a fuel spill occur.

18.8 Communications

Boom Microphones **(not AFM)** **If Operative:**	Must be used below 18,000 Feet
HF Radios ▲ **Prohibited Use HF Frequencies:**	♦ 11.133 MHz ♦ 22.683 MHz ♦ 29.489 MHz ♦ 22.434 MHz ♦ 22.766 MHz ♦ 29.490 MHz
Use of One HF radio for Transmission ▲**:**	Deselect other HF radio on all audio select panels (prevents audio interference)
Use during refueling:	CAUTION: Do not use
Broadband Comm Switch ▲ **During Ground De-Icing:**	If installed, Broadband Comm Switch must be OFF.

18.9 Electrical

Circuit Breaker Resets **Pulling and Resetting:**	Not authorized unless specified in a procedure.

18.10 Engines

Icing Conditions ▲ **Definition:**	OAT is 3°C or below
Ground Operations ▲ **When a Run Up is Required:**	In icing conditions when OAT is 3°C or below
Icing Conditions **Run Up Parameters:**	Momentary, to minimum 50% N_1 at intervals not to exceed 60 minutes.
200ER ***Freezing Fog*** **OAT 0°C to -6°C:**	Run up engines to 50% N_1 for 1 minute every 45 minutes taxi time
(Source: AD 2008-02-05) **OAT -7°C to -13°C:**	Run up engines to 59% N_1 for 1 minute every 45 minutes taxi time
Colder than -13°C:	No runup procedure—engines must be manually de-iced

Ground Operation in Freezing Fog 200ER

(AD 2008-02-05) **Visibility < 300m (985') *and* OAT 0 to -6°C:**	Run up engines to 50% N1 for 1 minute every 45 minutes taxi time
or **OAT -7 to -13°C *and* taxi time > 45 minutes:**	Run up the engines to 59% N1 for 1 minute every 45 minutes taxi time
or **OAT < -13°C *and* taxi time > 45 minutes:**	No run-up procedure; the engines must be manually de-iced.
Any temperature, if takeoff within 60 minutes total taxi time:	Regardless of temperature, If the core ice shedding procedure (above) is not accomplished within 45 minutes total taxi time in freezing fog w/ visibility ≤300m (985') But takeoff can be achieved within 60 minutes total taxi time in those conditions: ♦ Takeoff is permitted ***and*** ♦ A borescope inspection is required within 10 flights.
Ground Operation in Icing Conditions **Run Up:**	During ground operations (including taxi-in and taxi-out) in icing when the OAT is 3°C or below, the engine must be run up momentarily to a minimum of 50% N1 at intervals not to exceed 60 minutes.

Engine Type

	300ER	200 ER
	General Electric GE 90-115BL 115,000 pounds at Takeoff Thrust	Rolls Royce Trent 892 90,000 pounds at Takeoff Thrust

Temperature and Time Limits

Conditions	300ER Max EGT; (°C)	300ER Time Limit	200 ER Max EGT; (°C)	200 ER Time Limit
Starting:	750 Ground 825 Flight	No time limit	700	Momentary
Max Continuous Thrust (MCT) 300ER:	17,000' & Below			
	920	20 Seconds		
Takeoff 200 ER:			Takeoff	
			920	20 Seconds
			900	5 Minutes
Max Continuous Thrust (MCT):	All Altitudes		Max Continuous Thrust (MCT)	
	1050	No time limit	850	No time limit
	1090	5 minutes		

Engine Starter

Starter Motor Operation:

300 ER	200 ER
Starter air pressure during start is 30-45psi Up to 5 minutes (max) on, followed by 10 minutes off followed by 5 minutes (max) on **Note:** Allow to cool for 10 minutes before subsequent operation	Up to 3 minutes continuous operation then run down to 10% N3, followed by: A further 3 minutes continuous operation then run down to zero N3, followed by: Up to 1 minute continuous operation then run down to zero N3 and allow to cool for 30 minutes. ***OR*** Above operating procedures can be replaced by: Up to 6 1/2 minutes continuous operation then run down to zero N3 and allow to cool for 30 minutes.

Maximum Autostart Cycles: Autostart system may initiate a second auto-initiated attempt (200) or third attempt (300)

No AML entry unless EICAS message is generated after a failed autostart

Maximum RPM

	300ER	200 ER
N1:	110.5%	100.5%
N2:	121%	105.0%
N3:	N/A	102.5%

Oil System

	300ER	200 ER
Temperature ▲:	Maximum for 15 Minutes	Minimum for Engine Start
	143°C	-40°C
		Minimum for T/O Power
		50°C
	Max Continuous	
	132°C	191°C
Pressure ▲:	Minimum	Minimum at Idle
	10 psi [1]	25 psi
	[1]Temporary operation below 10 psi during negative G operation is limited to 15 seconds maximum	Min to Max Continous Thrust
		50 psi
	Max Continuous	
	132°C	191°C
Quantity:	21 Quarts	16 Quarts
Post-flight Contact maintenance if below:	22 Quarts	18 Quarts

AOM Secure Checklist, 14f.4.2 Expanded Checklist

Reverse Thrust ▲

Ground:	Use for ground operation only
In-flight:	Intentional selection of reverse thrust in flight is prohibited
Powerbacks:	Backing the airplane with use of reverse thrust is prohibited
Thrust lever movement from reverse to forward:	Position thrust reverse levers full down (forward thrust) only after engines have decelerated to reverse idle
After Rejected Takeoff (RTO):	300ER: — / 200 ER: If thrust reversers were deployed—T/O prohibited until maintenance complete.

Engine Limit Display Markings ▲

Maximum & Minimum: RED

Cautionary: AMBER

Single Engine Taxi

Not Authorized when: (WIPT)
- *Wet or slippery taxiways*
- *Icy conditions*
- *Prohibited by Ops Advisory pages*

Not Recommended for:
- *Tight turns/confined areas expected*
- 300ER (always) — AOM 5.2.1 Single Engine Taxi

APU Use: Not required for SE taxi — AOM 5.2.1 Single Engine Taxi

Reversers

Inspection Requirement: 200 Following any rejected takeoff in which thrust reverser(s) were deployed, maintenance must *determine* if an inspection is required *prior to subsequent takeoff*. QRH Brake Temp, p. 14.6

AOM 6.4.7 Maintenance Inspection

LIMITATIONS section & ASM: Supplementary Procedures 21f.1.2, p. 21f-8

General Preflight Flow, 3f.2.3 Expanded Flow, p. 3f-14

18.11 Flight Controls

Flight Control Mode ▲ — **Authorized T/O Mode:** Takeoff is permitted only in the **NORMAL** flight control mode

Flaps — See AOM 18.2.1 Altitude Limits and Maximum Flight Operating Speeds and page 6 of this guide.

Flight Maneuvering Load Acceleration Limits ▲
- **Flaps Up:** +2.5 to -1.0G
- **Flaps Down [1]:** +2.0 to 0.0 G

[1] With flaps 25 and 30, positive limits vary linearly from +2.0g at maximum landing weight to +1.5g at maximum takeoff weight.

Flight Controls, General — **Manual Inputs:** Avoid rapid & large alternating control inputs, especially in combination with large changes in pitch, roll, or yaw (e.g. large side slip angles) as they may result in structural failure at any speed, including below V_A.

18.12 Flight Instruments

Ground Maneuver Camera System 300 ER — **Use:** Should not be used during takeoff approach, and landing

RNAV (RNP) — **Minimum Required Navigation Performance (RNP)**
- **RNAV (RNP) Approaches:** ≤0.3 NM (or as published)
- **For RNAV (RNP) Missed Approaches:** No Restriction
- **Maximum Speed on Final Approach Segment:** 180 Knots
- **Minimum RNAV RNP Authorized:** 0.11 NM

Note: *To get to / change RNP in FMC: Select CDU* PROG *page, then button 6R*

Reference: AOM, 10.6.3 Approach Requirements

18.13 Flight Management, Navigation

QFE Operations — **Altimetry Use of QFE / QNH:** The use of QFE instead of QNH is not authorized.

Inertial Reference System (IRS)

Maximum Ground Alignment Latitude ▲ **N & S:** 78° 14.75'
Air Data Inertial Reference Unit (ADIRU) alignment must not be attempted at latitudes greater than 78°, 14.75 minutes.

Radio Altimeter Flight Restrictions — **Reference: AD 2023-10-02**
All aircraft have been modified to be radio altimeter tolerant, therefore the restrictions contained in FAA AD 2023-10-02 do not apply.

Radio Altimeter 5G C-Band Interferenc — **Approach, Landing and Go-Around: Reference: AD 2023-12-05**
All aircraft have been modified to be radio altimeter tolerant, therefore the restrictions contained in FAA AD 2023-10-02 do not apply.

RNAV Use for SIDs and STARs — **Building:** May not be built—must be loaded from current NAV data base.
AOM 10.7.4 RNAV Visual and FMS/CVFP Visual Approaches
AOM 3.5.2 FMS Preflight, p. 3-19

Holding—Maximum Speed - FAA
- **≤6000':** 200 KIAS — Reduce to holding speed 3 minutes prior to estimated initial time limit over holding fix.
- **>6000 through 14,000':** 230 KIAS — *May be restricted to 210 when on a low altitude chart*
- **>14,000':** 265 KIAS — FOM Arrival > General > Holding Speed 8g.3.2

Holding—Maximum Speed - ICAO
- **≤ 14,000':** 230 KIAS
- **>14.000 through 20,000':** 240 KIAS
- **20,000 through 34,000':** 265 KIAS
- **>34,000':** 0.83 Mach

Exceeding maximum holding airspeed may result in aircraft excursions beyond the holding pattern protected airspace.

NOTE: Advise ATC immediately if unable to comply with the maximum holding airspeed and request an alternate clearance.

Operational Requirements. For aircraft with an operational requirement (e.g., turbulence, icing, etc.), it may be necessary to hold at airspeeds in excess of the maximum speeds.

Non-Radar Environment. In a non-radar environment, advise ATC you cannot accept the assigned hold.240 KIAS May be restricted to 210 when on a low altitude chart. FOM Arrival > General > Holding Speed 8g.3.3

Approach Categories

	Category	Approach Speeds (KIAS
200ER :	C	121-140
300ER :	D	141-165

AOM 10.1.3 Aircraft Approach Category

18.14 Fuel

Fuel Quantity Indication System (FQIS) ▲

Source: AD 2020-11-11 — **Door Cycling Requirement:** Flight crew must verify the aircraft refueling station door was cycled.
When fuel is added to or removed from the center fuel tank, the Flight Crew must acknowledge they have received notification that the aircraft refueling station door has been cycled in accordance with Fuel Quantity Indicating System Refueling Door Cycling Procedure. See AOM 21k.1.1

Fuel Scavenge Sustem

Effectivity: 7AA-7CA Only
- **Applies when:** Center tank fuel is required for the mission
- **Flight Plan Notations:** An additional 700 lbs.(320 kg) of reserve fuel must be added to the center tank fuel load.
- **Source: AD 2019-16-13**

Must appear on the flight plan/dispatch release when:
- FKY > RLS FUEL is greater than 117,000 pounds.
- FOS > TOTAL fuel is greater than 117,000 pounds.
- There is center tank fuel (up to 3000 pounds) with less than full main tanks.

Approved Fuels
- **Approved for Use:** Jet A, Jet A-1, JP-5, and JP-8
- **Prohibited Fuels:** JP-4 and Jet B
- **Other Fuel Usage:** **Note:** Must be coordinated with MOC

Density ▲ — **Must be between:** 6.1 and 7.1 pounds per gallon

Fuel Temperature ▲ **Maximum:**	49°C (120°F)
Minimum Prior to Takeoff:	Must not be less than the greater of ♦ -40°C, *OR* ♦ 3°C above the freezing point of the fuel being used
In Flight:	Must be maintained at least 3°C above the freezing point of the fuel used
Use of Fuel System Icing Inhibitor Additives:	**Note:** Does not change the minimum fuel tank temperature limit.
Fuel Management ▲ **Fuel Loading:**	♦ Tanks may be loaded individually, simultaneously, or in any sequence ♦ Main Tanks must be loaded equally to desired fuel quantity or until full
Center Tank If Loading Fuel in Center Tank:	Main tanks must be *scheduled* to be full **Note:** *Center tank may contain up to 3,000 pounds of fuel with less than full main tanks provided the weight of the fuel in the center tank plus the actual zero fuel weight does not exceed the Maximum Zero Fuel Weight, and the Center of Gravity limits are observed.*

Usable Fuel Tank Max Quantities

	200ER	300ER
Main L or R:	64,100 lbs	69,000 lbs
Center:	174,900 lbs	182,800 lbs

1. *Based on density of 6.7 lbs. per gallon.*
2. *Total fuel at standard density is* 300ER *320,800;* 200ER *303,100 lbs.*

Fuel Crossfeed **Min Fuel Crossfeed Valve Use:**	For minimum fuel operation, a crossfeed valve is opened.
Correcting Fuel Imbalance:	Open a crossfeed valve and turn off the Fuel Pump Switches for the fuel tank that has the lowest quantity.
Fuel Imbalance ▲ **Between Main Tanks, for Taxi, Takeoff & Landing With Main Tank Fuel**	
> 123,000 300ER or 114,000 200ER lbs:	3,000 pounds *Average— 61,500* 300ER*, 57,000* 200ER *lbs per tank.*
≤ 90,000 300ER or 50,000 200ER lbs:	4,000 pounds *Average— 45,000* 300ER*, 25,000* 200ER *lbs per tank.*
Between ranges above for 300ER or 200ER respectively:	Linear interpolation between 3,000 and 4,000 pounds.

18.15 Landing Gear & Brakes

Gear-Related Speed Limits ▲	See AOM 18.2.2, Speed Limits, or page 6 of this guide.
Brakes **Inspection Required:**	Following any abort in which maximum (RTO) braking was used or any high speed rejected takeoff (>80 Knots) QRH Brake Temp, p. 14-6 1.12 Brake cooling charts may also generate an inspection requirement QRH Landing Gear 14.2, p. 3f-15
Tire Pressure Minimums **Nose Gear:**	190 psi
Main Gear:	200 psi AOM 3f.2.3, p. 3.f-15, Pre-Flight

18.16 Warning Systems

TCAS **Deviation Authorization** ▲**:**	Pilots are authorized to deviate from their current ATC clearance to the extent necessary to comply with a TCAS RA
Navigation Display Reliance for Maneuvers:	Maneuvers must <u>not</u> be based solely on traffic information presented on the navigation display.
Compliance with TCAS Resolution Advisories:	Required, unless the pilot-in-command determines that doing so would jeopardize the safe operation of the flight.
Enhanced Ground Proximity Warning System (EGPWS) ▲ **Navigation Using Terrain/Obstacle Awareness Display:**	Navigation must <u>not</u> be predicated upon the use of the Terrain/Obstacle Awareness Display.
Deviation Authorization:	Pilots are authorized to deviate from their current ATC clearance to the extent necessary to comply with an EGPWS warning.
Use of Look-Ahead Terrain Alerting, and Terrain Display Functions:	Prohibited within 15 nm of takeoff, approach, or landing at an airport or runway not contained in the GPWS terrain database.
Airports Included in GPS Database:	**Note**: All authorized airports (regular, provisional, refueling, alternate, and designated emergency airports) are contained in the GPS airport database.
Notification of Non-GPS Database Airports:	**Note**: For runwavs/airports not contained in the installed EGPWS terrain database, crews will be notified via Company Alert Message (F-4) and/or flight plan/dispatch release remarks.

Inflight Maneuvers

Note: *Only selected maneuvers are reproduced here for study. Notes are abbreviated for brevity and ease of study. Refer to AOM Insight Maneuvers section for other procedures and full text of notes. Procedures are taken from the AOM introduction to the Maneuvers chapter, which states, "**This chapter contains inflight maneuver information that may occur in multiple phases of flight.**"*

Approach to Stall or Stall Recovery

All recoveries from approach to stall should be done as if an actual stall has occurred. Immediately do the following at the first indication of stall (buffet or stick shaker).

Reduction of angle of attack is the most important action to recovery from an impending stall or full stall.

WARNING

*Excessive use of pitch trim or rudder may aggravate the condition, or may result in loss of control or in high structural loads.

NOTE

Do not use flight director commands during the recovery. If autopilot response is not acceptable, it should be disengaged. If autothrottle response is not acceptable, it should be disconnected.

NOTE

If autopilot response is not acceptable, it should be disengaged. If autothrottle response is not acceptable, it should be disconnected.

(AOM, Inflight Maneuvers 17.1):

Pilot Flying	Pilot Monitoring
Recognize and confirm the developing situation.	
First indication of stall (buffet or stick shaker)	
"My aircraft."	
Initiate the recovery: — Smoothly apply nose down elevator to reduce the angle of attack until buffet or stick shaker stops. Nose down stabilizer trim may be needed.*	Monitor altitude and airspeed Verify all required actions have been done
Continue the recovery: — Roll in the shortest direction to wings level if needed** — Advance thrust levers as needed — Retract the speedbrakes — Do not change gear or flap configuration, except • During liftoff, if flaps are up, call for flaps 1	Call out — Any omissions — Trend toward terrain contact Set the FLAP lever as directed
Complete the recovery: — Check airspeed and adjust thrust as needed — Establish pitch attitude — Return to the desired flight path Re-engage the autopilot and autothrottle if desired	Monitor altitude and airspeed Verify all required actions have been completed Call out — Any omissions — Trend toward terrain contact

Upset Recovery
Nose High Recovery Actions & Callouts

WARNING

* Excessive use of pitch trim or rudder may aggravate an unusual attitude or may result in loss of control and/or high structural loads.

(AOM, Inflight Maneuvers 17.2.1)

Pilot Flying	Pilot Monitoring
Recognize and confirm the developing situation	
First indication of nose high upset	
"My aircraft."	
— Disengage autopilot — Disconnect autothrottle — Apply as much as full nose down elevator — *Apply appropriate nose down stabilizer trim — Reduce thrust — *Roll (adjust bank angle) to obtain a nose down pitch rate — Complete the recovery: • When approaching the horizon, roll to wings level • Check airspeed and adjust thrust • Establish pitch attitude	Call out — Any omissions — Attitude, airspeed and altitude throughout the recovery
	Verify all required actions have been completed

Upset Recovery
Nose Low Recovery Actions & Callouts

WARNING

* Excessive use of pitch trim or rudder can aggravate an upset, result in loss of control, or result in high structural loads.

(AOM, Inflight Maneuvers 17.2.2)

Pilot Flying	Pilot Monitoring
Recognize and confirm the developing situation	
First indication of nose low upset	
"My aircraft."	
— Disengage autopilot — Disconnect autothrottle — Recover: • Recover from stall, if needed • Roll in the shortest direction to wings level (unload and roll if bank angle is more than 90°)* — Recover to level flight: • Apply nose up elevator • Apply nose up trim, if needed * • Adjust thrust and drag, if needed	Call out — Any omissions — Attitude, airspeed and altitude throughout the recovery
	Verify all required actions have been completed

Windshear or Microburst Escape

[1] If go-around is not available, disconnect autopilot and autothrottle and manually fly the aircraft.
[2] Severe windshear may exceed the performance capability of the AFDS.
[3] Maximum thrust can be obtained by advancing the thrust levers full forward if the EECs are in the normal mode. If terrain contact is imminent, advance thrust levers full forward.

(AOM Inflight Maneuvers 17.3.1)

	Pilot Flying	Pilot Monitoring
When encountering a windshear		
	Call **"Escape"**	
	"My aircraft."	
Auto	– Push either TO/GA switch[1] – Ensure TO/GA mode annunciation and GA thrust – Monitor system performance **WARNING:** Be prepared to disconnect the autopilot and autothrottle and manually fly the aircraft.[2]	• Ensure all required actions are completed
Manual	**Simultaneously:** – Disconnect the autopilot – Push either go-around switch – Aggressively apply maximum thrust[3]	Call out • any omissions • altitude and trend information based on radio altimeter (e.g., "300 feet descending" or "400 feet climbing")
Manual	– Disconnect the autothrottle – Roll wings level and rotate toward an initial pitch attitude of 15° – Follow TO/GA guidance (if available) without exceeding pitch limit indication (PLI)	
	Ensure the speed brakes are retracted	
	Do not: • change gear/flap configuration • attempt to regain lost airspeed *until* windshear is no longer a factor	
After escape is successful		
	• Resume normal flight • Retract gear and flaps as required	• Issue PIREP to ATC

Windshear Alerts: Takeoff

[1] Inhibited from 80 knots to 400 feet RA.
[2] Inhibited from 100 knots to 50 feet RA.
[3] Inhibited until rotation.

(AOM, Inflight Maneuvers 17.3.3)

Alert	Prior to V1	At or Above V1
Caution[1] "Monitor radar display."	Delay/reject the takeoff	• Maneuver as required to avoid windshear • Use maximum thrust
Warning "Windshear ahead. Windshear ahead." [2] or "Windshear. Windshear. Windshear." [3]	Delay/reject the takeoff	Perform the Windshear Escape Maneuver
Unacceptable Flight Deviations	Reject the takeoff	• Perform the Windshear Escape Maneuver • At V_R, rotate normally to 15° no later than 2000 feet runway remaining

Windshear Alerts: Approach

[1] **Inhibited below 400 feet RA.**
[2] Inhibited below 50 feet RA.

(AOM, Inflight Maneuvers 17.3.3)

Alert/Aural	During Approach
Caution [1] "Windshear." or "Monitor radar display."	– Execute a normal go-around and maneuver as necessary to avoid windshear **OR** – Continue the approach at captain's discretion if able to avoid windshear[1]
Warning [2] "Go around. Windshear ahead."	– Perform at pilot's discretion either: • A normal go-around or • The windshear escape maneuver
Warning "Windshear. Windshear. Windshear."	Perform the windshear escape maneuver
Unacceptable Flight Deviations	

Severe Weather / Windshear Procedures

From: AOM 17.3.2 Severe Weather Windshear Flow Chart

1. **Follow standard operating procedures:**
   – **Takeoff**: Minimize pitch attitude reductions
   – **Approach**: Avoid large thrust reductions
   – Know normal attitude and performance
2. **If predictive W/S caution or warning is received:**
   – **Aligned** for takeoff: Delay takeoff
   – **Prior to V_1** : Reject takeoff
   – **Above V_1** : Perform W/S Escape Maneuver
3. **Predictive W/S warning is received during approach ("GO-AROUND, WINDSHEAR AHEAD" aural):**
   – **Perform** W/S Escape Maneuver or, at pilot's discretion, perform a normal go-around
4. **Windshear encountered inflight:**
   – **Perform** W/S Escape Maneuver
   – **Report** encounter

Ground Proximity WARNING Actions & Callouts

200 ER – activation of the **"PULL UP"**, or **"TERRAIN TERRAIN PULL UP"** warning

300ER – activation of the **"PULL UP"**, or **"OBSTACLE OBSTACLE PULL UP"**, or **"TERRAIN TERRAIN PULL UP"** warning

Other situations resulting in unacceptable flight toward terrain

Note

Aft control column force increases as the airspeed decreases. In all cases, the pitch attitude that results in intermittent stick shaker or initial buffet is the upper pitch attitude limit. Flight at intermittent stick shaker may be required to obtain positive terrain separation. Smooth, steady control will avoid a pitch attitude overshoot and stall.

Do not use flight director commands.

If positive visual verification is made that no obstacle or terrain hazard exists when flying under daylight VMC conditions prior to an obstacle or terrain warning, the alert may be regarded as cautionary and the approach may be continued.

(AOM, Inflight Maneuvers 17.4.1)

Pilot Flying	Pilot Monitoring
Ground proximity *warning* alert	
"My aircraft."	
Immediately and simultaneously: – Disengage autopilot – Disconnect autothrottles – Aggressively apply maximum* thrust – Simultaneously roll wings level and rotate to an initial pitch attitude of 20° – Retract speedbrakes – If terrain remains a threat, continue rotation up to the pitch limit indicator or stick shaker or initial buffet – Do not change gear or flap configuration until terrain separation is assured – Monitor radio altimeter for sustained or increasing terrain separation – When clear of the terrain, slowly decrease pitch attitude and accelerate	– Assure maximum* thrust – Verify all required actions have been completed Call out any omissions Monitor vertical speed and altitude (radio altitude for terrain clearance and barometric altitude for a minimum safe altitude) Call out: – Any omissions – Any trend toward terrain contact **Note** * Maximum thrust can be obtained by advancing the thrust levers to the takeoff or go-around limit. On aircraft with EECs operating normally, the pilot may advance the thrust levers full forward. If terrain contact is imminent, advance thrust levers full forward.

Ground Proximity CAUTION Actions & Callouts

300ER Accomplish the following maneuver for any of these aural alerts:

- CAUTION OBSTACLE
- CAUTION TERRAIN
- TERRAIN
- DON'T SINK
- GLIDESLOPE
- SINK RATE
- TOO LOW FLAPS
- TOO LOW GEAR
- TOO LOW TERRAIN
- BANK ANGLE

(AOM, Inflight Maneuvers 17.4.1)

Pilot Flying	Pilot Monitoring
Ground proximity *caution* aural alert activated	
Correct the flight path or the aircraft configuration	

Note

If a terrain caution occurs when flying under daylight VMC, and positive visual verification is made that no obstacle or terrain hazard exists, the alert may be regarded as cautionary and the approach may be continued.

Moderate to Severe Turbulence Encounter

Speed..Turbulent Air Penetration
- 270 KIAS below 25,000 feet
- 280 KIAS/.82 Mach (whichever is lower) at and above 25,000 feet. Maintain a minimum speed of 15 knots above the minimum maneuvering speed (amber band) at all altitudes when the airspeed is below 0.82 Mach

Flap extension.................................Delay as long as possible
The aircraft can withstand higher gust loads in the clean configuration

Seat Belt Signs selector ....ON

Make PA
Refer to Flight Manual Pt 1 - Section 11 - Non-Routine PAs Typical Examples - Turbulence.

Autothrottles...As required
In severe turbulence during cruise, it may be necessary to disconnect the autothrottles to prevent excessive thrust changes. Thrust setting guidance is available on EICAS when VNAV is engaged. Set EPR/N1 at or slightly above the magenta VNAV target EPR/N1 indication. Change thrust setting only if required to modify an unacceptable speed trend.

Autopilot...Monitor
Use the autopilot in turbulence. Closely monitor autopilot operation and be prepared to disconnect the autopilot only if the aircraft does not maintain an acceptable attitude. If the autopilot disconnects, the pilot should smoothly take control and stabilize the pitch attitude.

(AOM, Inflight Maneuvers 17.4.1)

Traffic Advisory (TA) Actions & Callouts

WARNING

Comply with RA if there is a conflict between RA and air traffic control.

Once an RA has been issued, safe separation can be compromised if current vertical speed is changed, except as necessary to comply with the RA. This is because TCAS II-to-TCAS II coordination may be in progress with the intruder aircraft, and any change in vertical speed that does not comply with the RA may negate the effectiveness of the other aircraft's compliance with the RA.

(AOM, Inflight Maneuvers 17.5.1)

Pilot Flying	Pilot Monitoring
TCAS traffic advisory (TA) occurs	
Look for traffic using traffic display as a guide	
Call out any conflicting traffic	
If traffic is sighted, maneuver if needed	

Note

See numerous additional notes in Operating Manual.

Resolution Advisory (RA) - Except a Climb in Landing Configuration Actions & Callouts

(AOM, Inflight Maneuvers 17.5.2)

Pilot Flying	Pilot Monitoring
TCAS resolution advisory (RA)	
"My aircraft."	
– If maneuvering is required, disengage the autopilot and disconnect the autothrottles – Smoothly adjust pitch and thrust to satisfy the RA command – Follow the planned lateral flight path unless visual contact with the conflicting traffic requires other action	
Attempt to establish visual contact.	
Call out any conflicting traffic	

Climb RA in Landing Configuration Actions & Callouts

Note: When responding to an RA, the aircraft should be maneuvered only as much as needed to satisfy the RA.

(AOM, Inflight Maneuvers 17.5.2)

Captain	First Officer
TCAS resolution advisory (RA)	
"My aircraft."	
– Disengage autopilot – Disconnect autothrottle – Advance thrust levers forward to ensure maximum thrust is attained and call for FLAPS 20 – Smoothly adjust pitch to satisfy the RA command – Follow planned lateral flight path unless visual contact with conflicting traffic requires other action	– Verify maximum thrust set. – On command, position flap lever to 20 detent
Verify a positive rate of climb on the altimeter	Verify a positive rate of climb on the altimeter
	"Positive Rate"
"Gear Up."	On command, Landing Gear Lever **UP**
Attempt to establish visual contact	
Call out any conflicting traffic	

Rejected Takeoff
After RTO Considerations

When stopped. perform procedures as required.

Prior to 80 knots, the takeoff should be rejected for any of the following:
- Activation of the master caution system
- System failure(s)
- Unusual noise or vibration
- Tire failure
- Abnormally slow acceleration
- Takeoff configuration warning
- Fire or fire warning
- Engine failure
- Predictive windshear warning
- If a side window opens
- If the aircraft is unsafe or unable to fly

Above 80 knots and prior to V1, the takeoff should be rejected for any of the following:
- Fire or fire warning
- Engine failure
- Predictive windshear warning or caution
- If the aircraft is unsafe or unable to fly

During the takeoff, the crewmember observing the non-normal situation will immediately call it out as clearly as possible.

See additional details in the AOM notes.

Note

If an engine failure occurs above TAC activation speed, TAC provides a rudder input, as needed, to help maintain directional control. TAC rudder input is available during forward thrust operations only, until speed is reduced below TAC activation speed. With TAC inoperative, the PF must make rudder inputs.

(AOM, Inflight Maneuvers 17.6.1)

Captain	First Officer
The captain decides to reject the takeoff	
"Reject, my aircraft."	
	If aircraft control is transferred **"Your aircraft."**
Immediately and simultaneously: – Close the thrust levers – Disengage the autothrottle – Apply maximum manual wheel brakes or verify operation of RTO autobrakes. If RTO autobrakes is selected, monitor system performance and apply manual wheel brakes if the EICAS advisory message **AUTOBRAKE** displays or deceleration is not adequate – Apply reverse thrust up to the maximum amount consistent with conditions – If speedbrakes do not deploy (or fail to remain deployed), manually deploy speedbrakes – Continue maximum braking until certain the aircraft will stop on the runway.	Verify: – Thrust levers closed – Autothrottle disengaged – Maximum brakes applied If EICAS advisory message AUTOBRAKES displays during rollout **"Autobrakes off."** Verify SPEEDBRAKE lever full aft **"Deployed."** or If speedbrakes do not deploy (or fail to remain deployed) **"No Speedbrakes"** If there is no REV indication(s) or the indication(s) stays amber **"No reverser left engine."** or **"No reverser right engine."** or **"No reversers."**
	Call out any omitted action items Apply forward pressure on the control column
At 80 knots	
	"80."
At 60 knots	
	"60."
When stopping is assured: – Start movement of the reverse thrust levers to reach the reverse idle detent before taxi speed. – After the engines are at reverse idle, mode the thrust levers to full down.	
Communicate the reject decision to the control tower and appropriate crew members as soon as practical – If needed, make PA, *"This is the captain. Remain seated. Remain seated. Remain seated."*	

Emergency Descent

CAUTION

Plan descent to remain above highest area terrain while altering flight path as needed for continued descent to 10,000 feet within PAX O2 time limits. Recommended high terrain descent profiles below may be adjusted as necessary to meet terrain clearance/PAX O2 requirements:

200 ER PAX 02 12 Mins

Level Off at 17,000 feet for 10 minutes (or until terrain clearance is assured) from the time passenger oxygen was deployed, then descend to 10,000 feet or below.

300ER PAX 02 22 Mins

Level Off at 25,000 feet for 19 minutes (or until terrain clearance is assured) from the time passenger oxygen was deployed, then descend to 10,000 feet or below.

If temperatures are colder than ISA, the altimeter will indicate higher than the actual altitude. The error may be significant and becomes extremely important when considering obstacle clearances in very cold temperatures.

In extremely cold weather conditions (approximately -30°C SAT and below), add 2000 feet to the initial level-off altitude until safe terrain separation is ensured.

(AOM, Inflight Maneuvers 17.6.1)
Also in Electronic checklist > Unannunciated Checklists & QRH > Miscellaneous

MCP Altitude .... **Select lower**
Note: If due to depressurization, where passenger oxygen has been actuated, plan to reach 10,000 feet within passenger oxygen time limits and continue flight at 10,000 feet or below, unless operational or terrain considerations dictate otherwise.
FLCH .... **Select**
Thrust Levers .... **Close**
Speedbrake handle .... **Full aft**
SEAT BELTS selector .... **ON**
Descend.................. **Straight ahead or initiate moderate bank (30° max)**

CAUTION

If structural integrity is in doubt, limit airspeed and avoid high maneuvering loads.

Target speed .... **Max VMO / MMO**
Transponder.................. **Code 7700**
(Unless voice communications with ATC are established.)
ATC.................. **Call**
Make PA .... **Use oxygen – fasten seat belts**
Advise Flight Attendants, via PA, when they can remove their oxygen masks (cabin altitude at or below 10,000 feet).
Approaching Level Off Altitude (1000 feet above):
Speedbrakes.................. **Retract**

Note

If depressurization has occurred and passenger oxygen has actuated, plan to reach 10,000 feet within passenger oxygen time limits and continue flight at 10,000 feet or below, unless operational or terrain considerations dictate otherwise.

Takeoff Engine Failure/Loss of Thrust After V_1 Actions & Callouts

See additional notes in the AOM.

[1] If the FD TAKEOFF mode is engaged at time of engine failure, the FD will command this speed schedule. However, do not follow the FD pitch command bar until safely airborne.
[2] Verify engine thrust is at takeoff thrust setting until aircraft is in clean configuration and obstacle clearance is assured.

(AOM, Inflight Maneuvers 17.7.1)

1 ENGINE

Pilot Flying	Pilot Monitoring
Engine failure or loss of thrust	
"My aircraft."	
Maintain directional control	
At V_R	
	"Rotate."
Rotate to takeoff attitude	
After liftoff	
Verify positive rate of climb on altimeter	Verify positive rate of climb on the altimeter
	"Positive Rate."
"Gear up."	On command, Landing Gear Lever
Use the following climb s Use the following climb speed to the TPS engine-out acceleration altitude: – **If engine failure occurs after V1, but not above V_2:** • Maintain minimum V_2 – **If engine failure occurs between V_2 and V_2 + 15:** • Maintain speed attained at time of failure – **If engine failure occurs at a speed greater than V_2 + 15:** • Increase pitch to reduce airspeed to V_2 + 15 [1]	Monitor engine and flight instruments [2]
At the TPS engine-out acceleration altitude	
Accelerate and command flap retraction on schedule	On command, retract flaps
Aircraft in clean configuration and VREF 30 + 80 (UP Bug)	
Call for the appropriate checklist(s)	
	Accomplish the appropriate checklist(s) on command

Engine Failure on Final Approach

(AOM 10.1.12)

See notes, Approach section of AOM

Normal Procedures Notes

Note: *This chapter contains notes and references to some of the more commonly needed numbers and page numbers buried throughout the normal procedures, abnormal and systems sections of AOM, and some information from the Flight Training Manual. To the extent practical, it is organized in the order these items are encountered in a normal flight profile.*

Type Approach & OM Page Reference	Lowest Visibility	AFDS Status Annunciation	Alert Height Required	Baro/Radio Bug*	Unique Call-Outs
Visual FOM 9g.3.1 Visual Approach or CVFP	3 Miles	N/A		Approach mins or OFF	
Non-ILS FOM 9g.3.4 through 9g.3.9, Non-ILS Approaches	As Published	N/A		Baro—MDA as DA or DDA (MDA + 50')	
CAT I FOM 9g.5.2 Category I	½ Mile / RVR 1800'	N/A		Baro—DA	
CAT II—Always with DH FOM 9g.5.3 Category II Airports with 300M published Vis Req:	1600 (500m) / 1200 (350m) 1000	**LAND 2** or **LAND 3**	No	Radio—DH	All CAT II and CAT III—*"300"*, based on RA **At DH**, *"Minimums,"* if voice callouts inop
CAT III FOM 9g.5.4 Category III	300 m	**LAND 2**	No	Radio—DH	
CAT III FOM Table 9g.7 CAT III (Fail Operational) Minimums	300 (75m)	**LAND 3**	No DH: Yes* DH: No	Radio—0' Radio—DH	

*Reference for Bug Settings: AOM 10.1.4 Planning, table on p. 10-7; See also Approach sections in FOM 9g.

Climb Speeds

Use ECON Cost-Index derived speed or the following:

Altitude	Gross Weight	Speed
Below 10,000 ft	All Weights	250 kts
Above 10,000 ft		310 kts / .84 m

Approach Types

LAND 2 (Fail Passive)
- Redundancy such that:
  - Single fault cannot cause significant deviation from flight path
  - Could be caused by single A/P inop
  - In the case of a failure, this mode will
    - Stabilize the flight path on the localizer and glide slope
    - NOT flare and land properly
    - Require a go-around if a failure is detected
- Cat II Authorized
- CAT IIIa Authorized with 50' DH
- CAT IIIb NOT Authorized

LAND 3 (Fail Operational)
- Redundancy such that:
  - Single fault cannot prevent the autopilot system from making an automatic landing
  - Must be done with three A/Ps
  - In the case of a failure, this mode will
    - Stabilize the flight path on the localizer and glide slope
    - Flare and land properly if the failure does not cause the **LAND 3** status to change
    - Require a go-around if a failure is detected
- Cat II, IIIa & IIIb Authorized

Final Approach Deviation Callouts

<u>Category</u> <u>Criteria</u>....................<u>Callout</u>
Airspeed ....+10 / -5 knots*..............***Airspeed***
Rate of Descent........<2000'—More than 2000 fpm......***Sink Rate***
..................<1000'—More than 1000 fpm*.....***Sink Rate***
Glide Slope ....> ½ Dot Deviation.........***Glide Slope***
Localizer......................> ⅓ Dot Deviation ....***Localizer***
Vertical Path Deviation.........Call out any......................***Path***
Lateral Track Deviation ....Call out any ....***Track***
VOR or NDB Deviation Call out any............***VOR*** or ***NDB***
Touchdown...First 1/3 of Runway Not to Exceed first 3000'*
* PM directs GO AROUND if PF not correcting
APPROACH, LANDING, GO-AROUND 15.4-6

Autoland Requirement

For the 777 Fleet, policy is that all CAT II and III approaches must end in
- Autolanding ***OR***
- Go-Around

Manual landing is <u>not authorized</u> when below CAT I weather minimums

DH and AH

Not flown at the same time
DH requires external visual reference before passing specific altitude
AH does not have this visual requirement—only electronic verification

Lavatory and Water Systems

Water shutoffs
- Each lavatory has one behind mirror
  - Mirror release button near wall under mirror
- Doors 2L and 4L have isolation zone shutoff valves
- Door 4L valve shuts off all water between doors 4 and 2
- Door 2L valve shuts off all water forward of doors 2L & R
- Aft Galley shutoff valve shuts off all water aft of doors 4L & R

Toilet Flush Shutoff
- Floor below toilet
- Can be used if toilet sticks in flush mode

Autobrakes

<u>**Must**</u> **be armed** (If operative) for any of the following:
- ♦ Runway length less than 7000 feet .... Setting 2 minimum
- ♦ RVR < 4000 or visibility < ¾ mile .... Setting 3 minimum
- ♦ Runway contaminated—standing water/snow/slush/ice
- ♦ Approach speed is increased above normal by procedure
- ♦ Braking conditions reported less than good
- ♦ CAT II and III landings ....Setting 3 minimum, if operable
- ♦ When minimum stopping distance is required ....Setting MAX
  - <u>Note</u> *MAX Auto Brake deceleration is slightly less than full manual braking*

<u>**Recommended**</u>: landing with gusty winds or crosswinds
Recommended settings
- **MAX AUTO**—When minimum-stopping distance is required
  - Deceleration rate is slightly less than that produced by full manual braking
- **3** or **4**—For wet / slippery runways or when landing rollout distance is limited
- **2**—For moderate deceleration

AOM 9.1.1 Preparation Items, p. 9-3

Auto Speedbrake System

Will be armed for all landings (If operative) ....AOM 10f.1.2
Do not land with the Speedbrakes deployed.........AOM 10f.1.2
Use on Approach
- Use with flaps over 5 causes buffet ....AOM 9.2.2
- Should be retracted by 1,000' AGL ....AOM 9.2.2
- Landing with thrust higher than idle may disable automatic deployment, causing a bounced landing ....AOM 12.3.1
- If speedbrakes deploy and airplane becomes airborne again,
  - They will retract while airborne ....AOM 12.3.1
  - Must then be manually redeployed ....AOM 12.3.1
- If speedbrakes NOT raised after touchdown, braking effectiveness reduces by up to 60%.............AOM 12.1.4

Takeoff Flap Selection

Authorized flap settings for takeoff are:

200 ER aircraft with TO thrust ....Flaps 5, 15, & 20
200 ER aircraft with TO1 or TO2 thrust............Flaps 5 & 15
300ER aircraft with TO thrust.....................Flaps 15 & 20
300ER aircraft with TO1 or TO2 thrust..................Flaps 15

Flaps 5, 15, & 20 are certified for takeoff on both aircraft, but not all combinations of flap setting / aircraft / thrust ratings are used by all operators.

Data for all of the authorized flap settings are not necessarily loaded in to TPS for all thrust ratings. Only data for the needed flap setting are loaded into TPS.

Contaminated Runway:

Flaps 15 (20 for 300ER) recommended for takeoffs on contaminated runways to minimize the takeoff speeds.

If flaps 15 (20 for 300ER) cannot be accommodated due to weight or conditions, coordinate another flap setting.

Landing Flap Selection

Use of flaps 25 is normally preferred:

- More fuel efficient
- Better noise abatement
- Reduced flap wear
- Improved maneuverability in high (gusty) wind conditions

Note: *The 777 is not certified for flaps 25 autoland.*

Use of flaps 30 is recommended when landing:

- Braking action reported less than good
- Tailwind
- Short runway (less than 8000 feet)
- Wet / contaminated runway
- When deemed prudent by the Captain

AOM 9.1.3 Landing Flap Selection

Approach Techniques

Initial Setup..**WARM**

Weather requirements..Check
Include crosswinds, visibility, published minimums

Approach...........................Load Approach from Database
....Set minimums
....Brief Approach

RNP...As Required

MCP—When cleared for the approach ....LNAV / VNAV
.... Set appropriate minimums when cleared for approach ***and*** on intercept heading ***or*** a portion of the approach

Captain Callouts...................................LOC / GS Intercept
....Passing FAF

Passing FAF Actions..**MAC-V**

Missed approach Altitude ....Set when 300' below MAP
Altimeters—Verify CA—F/O...................................± 100'
Clearance—Tower ....Call Passing FAF
V/S ....Set prior to MDA if end point not RWXX

Note: *This section compiled from instructor notes and various OM references; NOT procedural, but technique recommendations.*

Briefings and Restrictions

Flight Attendant Briefing—Prior to gate departure

Known delays

Maintenance—cabin or galley writeup and deferrals in AML

Ground—Short-taxi & safety demo considerations

Crew

- Number of F/As onboard and minimum crew requirements
  See paragraph 3g.10.2 Flight Attendants.
- Coordination of flightdeck crew meal service, if applicable
  Time for meals is left to the CA's discretion
  Cooperation should result in eating meals at desired time
- Need to minimize noise & activity near crew rest areas

Security

Details not included here for sensitivity concerns

Turbulence

- Weather/turbulence forecast for route: duration & severity
- Suspension of service at at flight attendant discretion
- Notification required action for unexpected turbulence
- Notify flightdeck of cabin situation & injuries if applicable
- If not seated at jumpseat with interphone, stay seated to avoid injury until instructed to resume duties by Captain

International

- Gen Decs/Customs Immigration forms (if required)
- N Atlantic flights north of 66°N over Greenland—request FA 1/purser review applicable FA Safety Manual information reference Supplemental O2 Unit ....FOM 3c.4.2

Use of Standard Thrust for takeoff is not authorized:

The following items require Maximum Takeoff Thrust at the TO rating:

- When using Improved Performance.
- If windshear is reported or expected.
- When Airport Ops Advisory (##-7) requires Maximum Thrust.
- When the runway is contaminated by standing water, slush, snow, or ice.
- Whenever takeoff data is used from the (Contaminated Runway Correction/MEL) CRC/MEL message for an MEL/CDL item.

However, the following items require Maximum Takeoff Thrust, but may be used with any thrust rating (TO2, TO1, or TO):

- If Takeoff Weight (TOW) exceeds Assumed Takeoff Weight (ATOW). However, a new TPS may be obtained which authorizes Standard Thrust at the Takeoff Weight.
- With MEL / CDL items containing a takeoff weight correction UNLESS takeoff data is automatically corrected by TPS (TPAS), or MEL/CDL item prohibits standard thrust.

Performance Manual, TPS, Introduction 10.4

AC to Pack Takeoff 300ER

Uses APU air to supply the left A/C pack

Use takeoff performance data based on A/C OFF

Applies to max weight, thrust, V-speeds, and assumed temperature..AOM Gen Ops, Performance 1p.2.8, p. 1p-12

Restricted Captain Requirements

Applies if

CA has < 100 hours after IOE in current aircraft type

Required 100 hours PIC in aircraft type may be reduced not to exceed 50% by substituting one landing for one hour of PIC time in aircraft type.

Restrictions *unless* FAA Exemption 5549 apply

- Chart increases required visibility .FOM Table 18.11, p.18-45
- Alternate Airport - Published minima need not be increased, but lowest allowable landing minima is 300 feet MDA/DH and one mile visibility or 4500 RVR / 1400 meters

Practical Effect of Exemption 5549:

Recognizes inherent safety of autopilot coupled approaches

Allows approaches using

- CAT II procedures to no lower CAT published minima or
- CAT I procedures to published CAT I MDA, no lower than RVR 1800'

See detailed charts ....FOM Table 18.11, p.18-45

Low Experience FO: <100 Hours (FAR 121.438)

When Applies—FO has <100 hours flight time as SIC in type aircraft being flown, ***and***

Captain (PIC) is not a qualified Check Airman,

PIC must make all takeoffs when:

All special requirement airports
Visibility in the latest WX report at or below 3/4 mile
RVR for the runway to be used is at or below 4000 feet
Runway has water, snow, slush, or similar conditions which may adversely affect A/C performance
Braking action is reported less than good
Crosswind component >15kts
Windshear reported in the vicinity of the airport
Any other condition PIC determines prudent ....FOM, 18-46

Flight Attendants—Emergency TEST Acronym

Use only if evacuation is anticipated

T - **T**ype of emergency; general description of the emergency

E - **E**vacuation is anticipated

S - **S**ignal: the following PA:
"This is the Captain. Evacuate. Evacuate. Evacuate."
Followed by turning on Evacuation **COMMAND.**

T - **T**ime: time to landing. ....QRH MISCELLANEOUS 0.24
See additional considerations regarding emergency evacuation in FOM ....FOM, 1N.6.3 Evacuation

Emergency PA to Passengers

Nature of emergency—General description of the situation

Time to landing

Safety vehicles may be present

Passenger cooperation with Flight Attendants

Reassure passengers ....QRH MISCELLANEOUS 0.24, 0.25

Cold Weather Operations

Cold Weather Operations Ref..........AOM 15.8 Cold Weather Ops

Icing Definition—Ground or Flight

OAT(Gnd) or TAT (Flt) < 10°C/50°F

AND Visible moisture present such as clouds, fog with visibility one mile or less, rain, snow, sleet, or ice crystals.

OR (Ground only) Temperature as above and
Any form of moisture is present (standing water, snow, slush etc.) ***which may be***
Ingested by the engines and freeze on engine inlets, nacelles or engine sensor probes

In Flight Use

Engine anti-ice must be **AUTO** or **ON** during all flight operations when icing conditions exist or are anticipated, except when the temperature is below-40°C OAT (SAT)

CAUTION: *Do not use engine anti-ice when TAT is above 10°C.*

Preflight

Advise Dispatch if TPS adjustments needed for snow, slush ice or standing water

Special Attention for ice in:
- Wheel wells, actuators & steering components
- Flight controls (flaps, slats, control tabs, etc.)
- Wing surfaces (Max 1/8 inch frost on bottom
- Pitot-Static components
- Engine Inlets
- Pack inlet/exit doors
- Fuel tank vents
- Pressure reg./relief valves
- APU Inlet

Caution: *Don't power up or move flight controls if ice could obstruct movement.*

Deicing

Packs and Bleeds (Engine & APU) OFF During and for 1 minute after deicing

APU--May be ON

Engines If operating, at idle power

Before T/O Checklist Must be completed in entirety following deicing

Lowest Operational Use Temperature (LOUT)

Lowest temperature at which de/anti-icing fluid flows off aircraft critical surfaces while maintaining the required freeze point buffer................AOM 15c.6 and fuel holdover tables

Start/Taxi

Caution: *If frozen contamination in intakes, must be removed by certified deicing personnel.*

Engine Start—Ensure normal N_1 during start
Prohibited during pushback on slippery ramp surfaces
Captain can decide not to start until pushback complete in any icing conditions

Oil Pressure May be high due to low OAT
- Operate at idle until in normal range
- May use above idle for crossbleed start if req.

Displays may require additional warm up time or initially appear slightly dimmer than normal

Engine Anti-Ice ON Immediately after start if icing conditions exist or anticipated before T/O

Flaps & Slats Up until just before takeoff *if*
Taxi route is through ice, snow, slush or standing water or
Freezing precipitation continues after deice/anti-ice

Flight Controls—Check for *full travel* of flight controls and *free movement* of trim

Run Engines Up for approximately 1 second to 50% N_1
Applies when OAT is 3°C, 37°F or less and anti ice is required
Done at least once every 60 minutes
AOM Cold Weather Operations 15.8.1

Freezing Fog Conditions (AD 2008-02-05)
Visibility less than 300 meters
Ice can build up on engine core
Shedding procedure—Momentary N1 run up to 50% every 45 minutes
Additional considerations:.............................AOM 15.8.1

Takeoff

If OAT is 3° C (37° F) or less static runup required
Minimum runup 50% N1
Confirm stable engine operation before starting takeoff roll

If runway contaminated, MAXIMUM thrust required

Takeoff is not authorized in:
- Hail
- Heavy Ice Pellets
- Moderate Freezing Rain ....De/Anti-Icing 87-3

Climb, Cruise, Descent

Engine anti-ice must be AUTO or **ON** during all flight operations when
- Icing conditions exist or are anticipated, except when
- SAT is below-40°C OAT (SAT).

CAUTION : Don't Operate Anti Ice above 10° TAT ..AOM 15.8.4 Taxi

Fan Ice Removal

If moderate to severe icing conditions encountered for prolonged periods, and
Fan icing suspected due to high engine vibration,
Fan blades must be cleared of any ice
300ER with N1 settings at or below 70%, every 15 minutes:
- Reduce thrust toward idle, then
- Increase to a minimum of 70% N1 for 10-30 seconds

200 ER
- Quickly reduce thrust to idle for 5 seconds then restore required thrust.
- If vibration persists, advance thrust lever to 90% N1 momentarily......AOM 15.8.6 Cold Weather, Inflight

Wing anti ice use in flight

Required when structural icing is present
Indicated by ice build-up on
- Flight deck window frames
- Windshield wiper arm
- Windshield center post
- Side windows

Primary method—Use automatic ice detection system, which acts as a deicer and allows ice to build up before turning wing anti ice on

Secondary method—Select **WING ANTI ICE** selector **ON** when wing icing is possible and use as an anti-icer

Ice Crystal Icing

Threat to engines due to ice ingestion and damage
See definition and characteristics ....AOM 15.3
Conducive condition recognition..................AOM AOM 15.3
Prevention—Avoid flying directly over large convective cells with red radar returns, even if no returns at A/C Altitude
AOM 15.3.2 Avoiding Ice Crystal Icing

Landing

Check Braking Conditions—ATIS, Tower, etc.

Runway Unfit for landing:
- Pools of water, (accumulations of more than ½ inch)
- Wet snow, or slush (accumulations of more than ½ inch)
- Chunks of ice or hardened snow
- "Braking action NIL" is reported ....FOM 11g.2.2

Slippery Runway Techniques
- Land on speed
- Touchdown at planned point (Firm is better than "grease job" on icy runway)
- Nose--Keep firmly on runway with elevator
- Auto Brakes and Spoilers—Use
- Max Rev Thrust ASAP after touchdown *buts*
  - -Watch for asymmetric reverser deployment
  - -Avoid excessive reverse on contaminated runway, which could lead to ice ingestion
  - -Don't come out of reverse at high RPM
- Use as much runway as needed to slow to taxi speed before turning off

Visual Perception problems possible with blowing snow—use caution

Taxi In & Parking

Do not retract flaps if:
Slush or Snow on runway *or*
Approach was in prolonged icing conditions
Call station for flap check, & deicing if req'd

Call Dispatch—Report runway as required

Freezing Fog Conditions (AD 2008-02-05)
Visibility less than 300 meters
Do run up momentary to 50% N1 within 5 minutes of parking aircraft in these conditions

Braking Action Report via ACARS

if A/C is parked overnight in freezing temperatures:
Draining Water System required
Draining Lavatory Tanks required unless anti-freeze is added
At off line stations with no AA maintenance, instructions available for station personnel or contract maintenance through Technical Services Group

Airplane General

Note: *This section contains notes on only selected "miscellaneous" components. They are the ones which, in the opinion of the author, are most necessary to know in time-critical situations.*

Lighting & Entertainment Systems

Passenger Cabin Signs

Fasten Seat Belt Signs—(10, 10, FOG)
In the **AUTO** position, on when:
10-A*ircraft below 10,000' MSL OR*
10-C*abin pressure > 10,000' pressure alt. OR*
F-F*lap handle not up OR*
O-P*assenger Oxygen on OR*
G-G*ear not up and locked*
In the **OFF** position, FSB signs ON when
O-P*assenger Oxygen on*

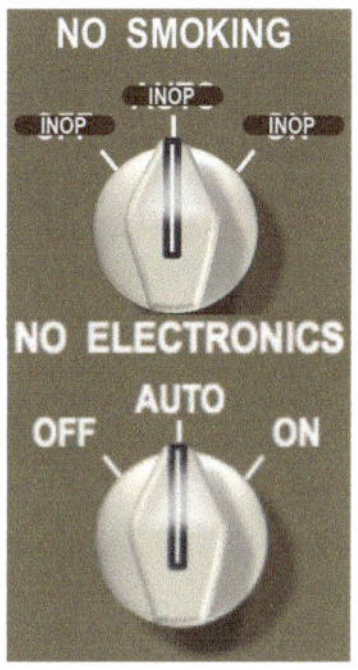

No Smoking Signs
On at all times
Chime sounds when moving switch between left and center positions
Chime ***does not*** sound when moving between center and right positions

Return to Seat Signs—Lavatories
ON whenever FSB signs are on ***except***
OFF when passenger oxygen is deployed
Found in lavatories

300ER No Electronics Selector
OFF—**ELECTRONIC DEVICES OFF** (EDO) signs not illuminated
ON—EDO signs are **ON**
AUTO—EDO signs on when
G—*Ground; On ground, flaps not up,* ***or***
L—*Low Altitude; In flight, cabin altitude < 10,000',* ***or***
O—*Oxygen; Passenger oxygen on* ....ASM, 1.40.8

Note: *if used for chime before takeoff, must be cycled quickly (<1 second) to prevent activating PED video*

Emergency Lighting

Controlled in cockpit by overhead panel EMER LIGHT Switch

Automatic operation occurs when:
System is armed ***and***
DC power fails or is turned off ***or***
An exterior door mode select lever is **ARMED** and the door is manually opened

In the latter case (door opened) exterior fuselage lighting at that door, and all interior emergency lighting comes on

Alternate control in main cabin on flight attendant panel at door 1L
Switch can be used to bypass cockpit switch
Will activate emergency lighting ***regardless of the cockpit switch position***

EICAS EMER LIGHTS message is displayed if
Emergency lights switch is not in the armed position ***or***
Emergency lights switch is armed, and the lighting is activated from the door 1L panel

Power for emergency lights
Remote batteries
Charged by airplane electrical system
Fully charged batteries provide at least 15 minutes of operation

Interior Emergency Lighting—Consists of:
♦ Door ♦ Aisle ♦ Cross-Aisle
♦ Escape Path ♦ Exit Lights ♦ Luminescent Exit Lights
Escape Path Lighting
Installed in arm rests
Designed to light path if smoke obscures lights above four feet from floor of cabin
Battery-powered exit lights at each cabin exit

Exterior Emergency Lighting
Lights aft of each door illuminate slide areas
Door 2L and 2R have lights built into the slide
Slides have a sharper angle from the fuselage than slides at other doors
Lighting designed to prevent "missing the slide"

Cockpit Lighting

Mag compass light is controlled with **GLARESHIELD** rheostat
Master brightness control—optimized when all controls with dots are adjusted to line up the dot and the marker line

Passenger Internet/Broadband Switch

Broadband Switch
OFF—disables two way broadband passenger communications.
ON—(guarded position) - enables two way broadband passenger communications.

Doors and Windows

Flight Deck Door

Reinforced doors installed throughout fleet
Door code located in secure file **RF 7700 CODE**
Additional information in FOM 7s.3.1 Flightdeck Door; AOM 20.2.3

Flight Deck Windows

Provide emergency exit from cockpit with escape ropes
Can be closed in flight if unlocked on takeoff
Recommended not to exceed V_{REF} + 80
May not be possible to close above 250 knots

Passenger Entry Doors

8 Doors—All translating, plug-type doors and operate the same
Numbered front to rear **1-4**, **L** or **R** (i.e. **1L**, **4R**, etc.)

Flight Lock
Each door automatically locks when airspeed is >80 knots
Spring-loaded solenoid held locked with electrical power
Unlock occurs with loss of electrical power

Slide Deployment
With doors armed, occurs when door opened from inside
Disarmed when door is opened from outside

Vent Panels
Connected to door handles
Insure cabin is depressurized before door opens
At high differential pressures, vent panel opens only partially until differential pressure is reduced
At very high differential pressure, panel can't be opened

Indicators
Power assist reservoir—if gas needle outside green zone, system is unusable
Slide / Raft gas bottle pressure gauge—if gas needle outside green zone, system is unusable
Girt Bar Indicator Flag viewing windows
Yellow—Door and slide / raft ***are*** armed
Black—Door and slide / raft ***are not*** armed

Cargo Doors

Forward and aft cargo doors— Open outward
Latched closed, not plug-type doors
Electrically operated

Bulk cargo door—Plug-type door
Manually opened
Opens inwardDoor Synoptic Display

Symbols used for Passenger Doors
Amber ....Door Open
Blank ....Door Closed
White ....Door Status Not Available

Symbols used for Cargo Doors
Amber ....Door Open
Blank ....Door Closed
White ....Door Status Not Available

Oxygen Systems

Flight Crew Oxygen System

One bottle supplies all cockpit masks (4)
Squeezing red mask release levers releases the mask from storage

Removing the mask:
Inflates the mask harness
Momentarily displays the yellow O2 flow indicator
Selects the mask microphone in the removed mask (deselecting boom mic)

Reselecting boom mic
Close left mask door ***and***
Press and release the **RESET/TEST** switch

Passenger Oxygen System

O_2 supplied from individual chemical O_2 generators
Located in passenger service units (PSUs) above passenger seats
Oxygen flows for a minimum of 12 minutes
300ER Oxygen flows for minimum of 22 minutes

Masks deployed:
Automatically—cabin altitude exceeds approx. 13,500' pressure
300ER Horn sounds in cockpit and crew rest area
Manually when cockpit overhead panel Passenger Oxygen Switch pressed to the **ON** position

Switch Indications
Blank—Normal position, system ready
Pressed—Signals PSU switches to drop masks
ON (*Amber*)—System is operating and masks have dropped

Emergency Equipment

Emergency Evacuation Signal System

Alerts Flight Attendants to evacuate passenger cabin

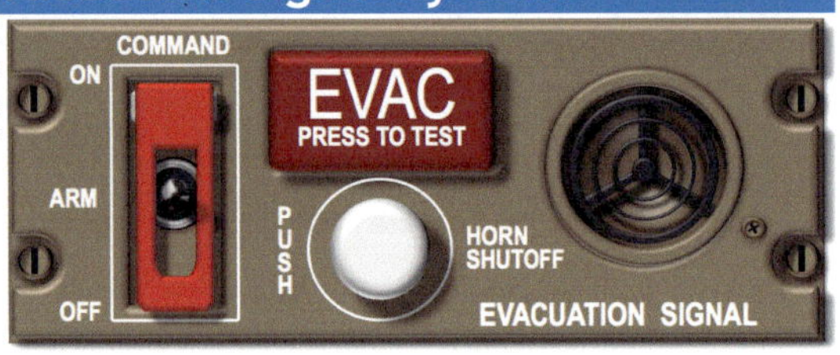

Switches located:
On the flight deck center console
At each flight attendant station

Operation
Flight Deck Evacuation Switch **ON** activates evacuation signal:
In cockpit ***and***
On flight attendant panels
Flight Deck Evacuation Switch **ARM**
Pressing flight attendant panel **EVAC COMMAND** switch activates evacuation command signals:
On the flight attendant panels
and
On the flight deck
Flight Deck Evacuation Switch **OFF**;
Pressing flight attendant panel **EVAC COMMAND** switch activates evacuation command signals:
On the flight deck ***ONLY***

Halon Fire Extinguishers

Halon Gas is toxic and inhalation is very dangerous
Works by displacing oxygen
Halon is heavier than other gases and will fall to the lower areas (floor vicinity)
Warning: *If a fire extinguisher is discharged in the flight deck area, all flight crew members must wear O_2 masks and use 100% oxygen with emergency selected*

Emergency Locator Transmitters

Installed in slide/raft containers at doors 1L and 4R (**300ER** 5R)

Transmit automatically when slide / rafts are deployed into the water

300ER **ELT Switch**
RESET—(spring-loaded) ends ELT transmission
ARMED—(guarded) Transmits emergency signal if activated by high deceleration forces
ON—transmits emergency locator signal.

Emergency Flashlights

Located at each flight attendant seat
Come on automatically
When removed from the holders
Cannot be turned off until returned to holder
Charging lights
Small LED flashes continuously if batteries serviceable
If no flashing light—battery replacement is called for

Enhanced Medical Kits and Automatic External Defibrillators (AEDs)

Locked into position with a bracket
Can be unlocked with cockpit door key
Neither is required for dispatch

200 ER Aircraft Principal Dimensions

Width	199' 11"	60.9m
Length	209' 1"	63.7m
Height	60' 5" to 61' 6"	18.4-18.8m
Main Gear Strut Width	36'	11m
300ER **Nose to Mains**	84'11"	*25.9m*

Dimensions below in diagram

For other dimensions, turn radius and taxi details see ASM 1.10.1 Principal Dimensions; 1.10.2 Turning Radius *or* AOM 5.3.4 Turn Radius; 5.3.5 Min Radius Turns / 180 Degree Turns

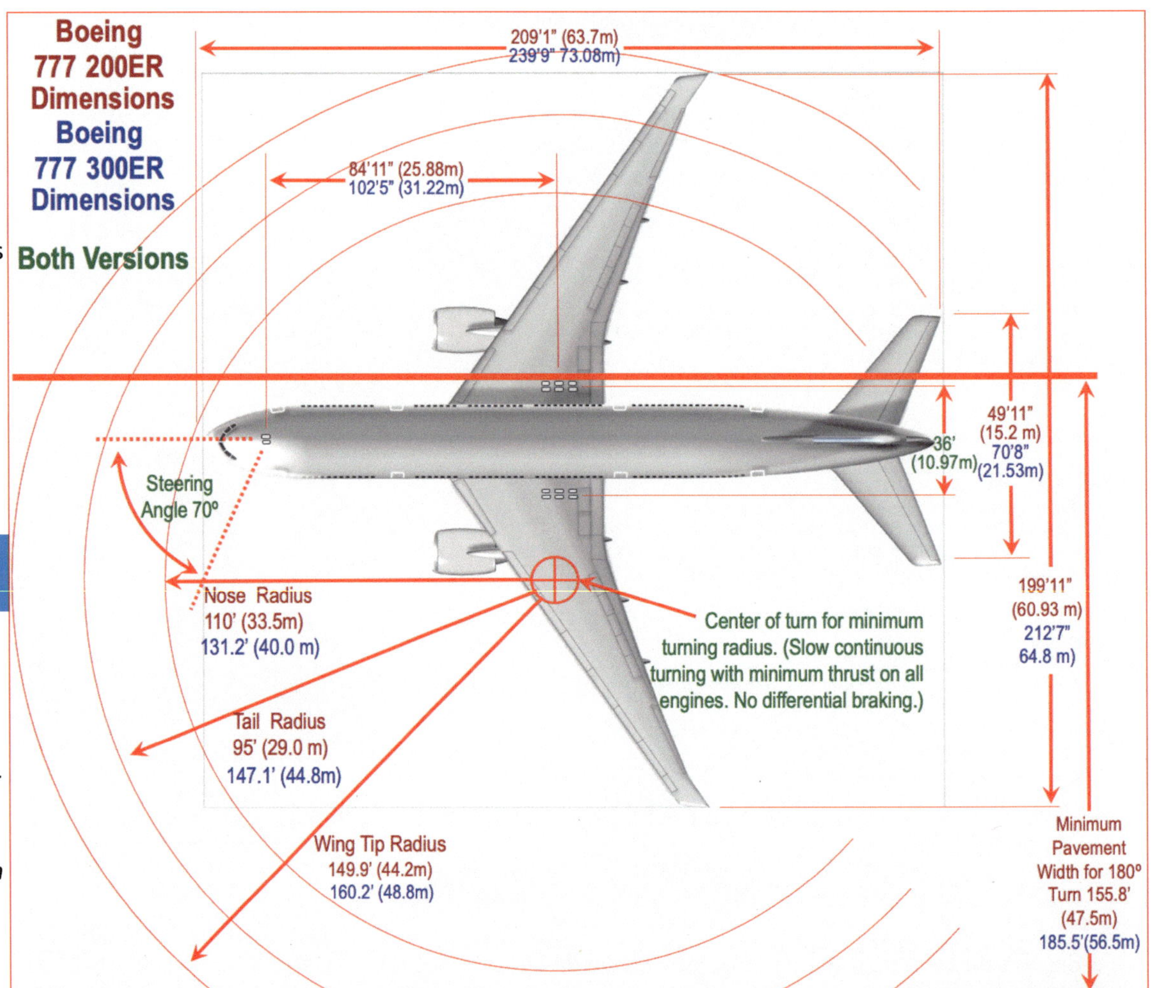

Side Silhouette drawing courtesy of ***Norebbo Stock Illustration and Design***, www.norebbo.com.

Emergency Equipment Location

200 ER

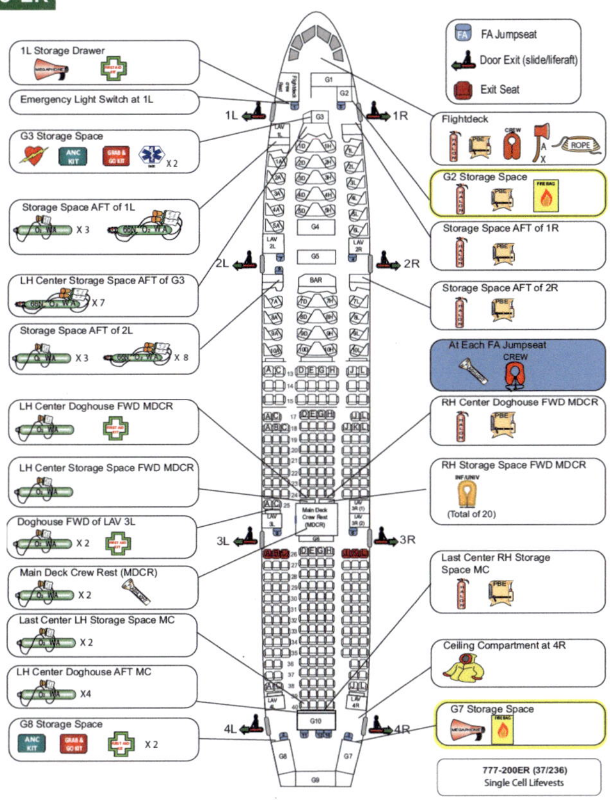

300ER

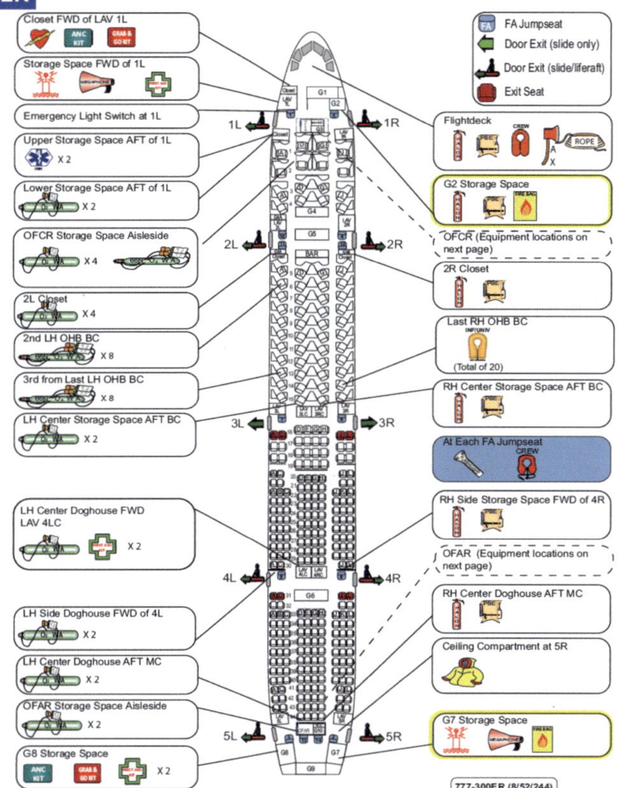

300ER Overhead Flight Crew Rest Area

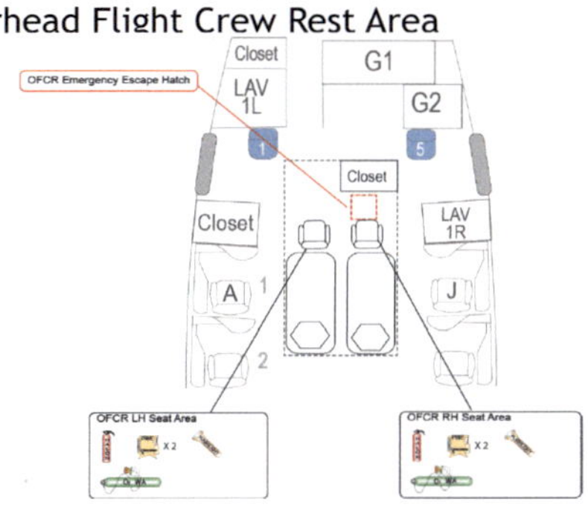

300ER Overhead FA Rest Area

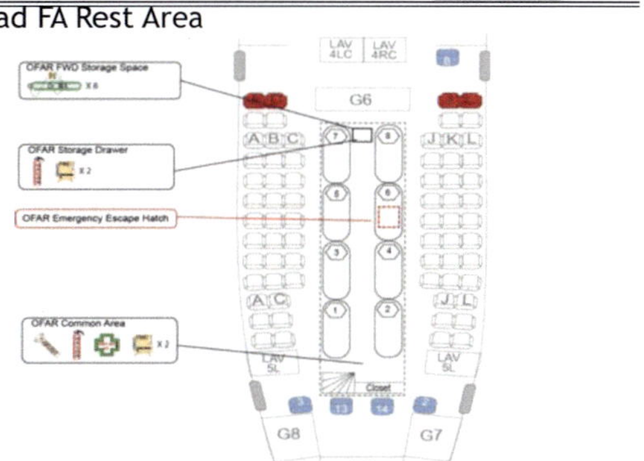

All Diagrams:
AOM Emergency Equipment, 20.3.2 Location Diagrams

Aircraft Configuration

	Code	First	Business	Premium Economy	Coach	Total
200ER	82	37	—	—	236	273
300ER	83	8	52	28	216	304

Other seating variations exist, but emergency equipment is located in the same positions.

Air Conditioning and Pressurization

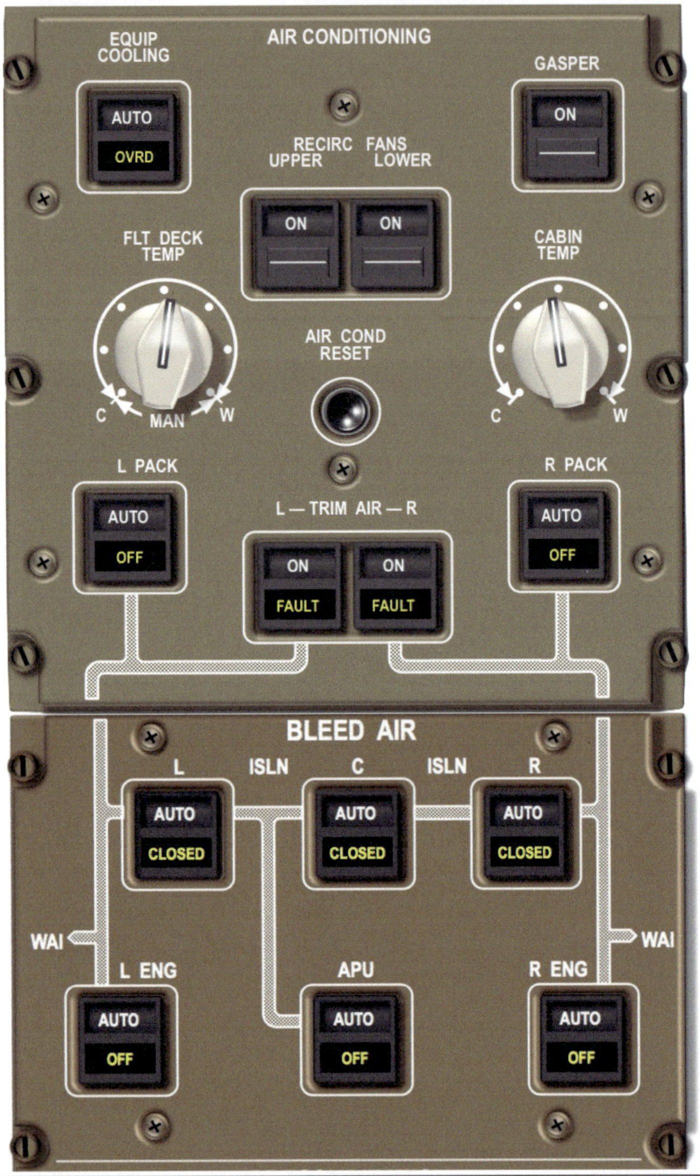

Bleed Air System

Engine Bleed Air

Valves are pressure actuated

Remain closed until sufficient pressure exists to cause forward flow

Engine bleed valves **close automatically:**

- ♦ During engine start
- ♦ Bleed source loss
- ♦ Bleed air overtemperature
- ♦ Bleed air overpressure
- ♦ Bleed air duct leak
- ♦ Engine fire switch pulled
- ♦ Ground air cart supplying air

APU Bleed Air Supply

Primary Uses

- ♦ Engine start
- ♦ Ground air conditioning
- ♦ Pressurization for no engine bleed takeoffs

300ER APU to Pack selection for takeoff accomplished through FMC

In flight use

Bleed air available up to 22,000 feet pressure altitude

Check valve in APU supply line prevents reverse flow back into APU

Ground Bleed Air Supply

Three connection points to bleed ducts

Check valves in supply line prevents reverse flow back into external sources

Duct Leak and Overheat Detection

If duct leak is detected, system automatically isolates the leak

Isolation process

May take one, two or three automatic steps

During process, bleed valves cycle and appropriate EICAS messages illuminate

When affected temperature cools

EICAS messages removed

System remains isolated

Switch **OFF** or **CLOSED** lights remain and show final status

Air Synoptic

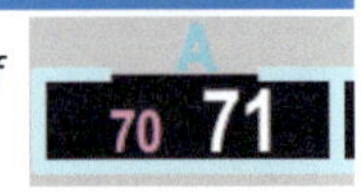

Selected temperatures are magenta

Not displayed when temperature control of associated zone is inoperative

Actual temperatures are white

Air Conditioning System

Packs

Two identical controllers on each pack

If a controller fails, the other is automatically selected

Pack output is increased during high demands ***OR***

To compensate for a failed pack or recirc fan

Pack output is reduced flow during high bleed air demand periods such as gear retraction ***AND***

During cruise flight to reduce fuel consumption

Ground operation

Dual pack operation is the normal mode

Fuel flow approximately the same with 1 or 2 packs operating

Noise level is higher with single pack operation

Non-Normal Operation

Protections are automatic for

Pack control

Fault detection

Overheat protection

If a significant fault is detected

Pack shuts down automatically

EICAS Message **PACK L** or **R** displays

Attempt to restore pack operation may be made by pressing Air Conditioning Reset switch

Pack Switches

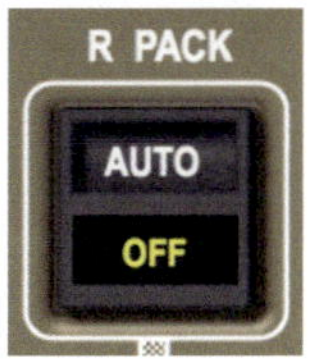

- **AUTO**—pack control is automatic
- **OFF** (**AUTO** not visible)—off
- **OFF**—pack flow control valve commanded closed
  - Illuminated when:
    - ♦ the PACK switch is pushed **OFF** ***OR***
    - ♦ Pack valve is failed closed, ***OR***
    - ♦ Pack or compressor outlet high temperature ***OR***
    - ♦ Bleed air pressure is inadequate, ***OR***
    - ♦ **300ER** R PACK valve is closed during APU to Pack Takeoff procedure

Standby Cooling Mode

Backup to normal mode

- EICAS Message **PACK MODE L** or **R** displays
- **STBY COOLING** (*Amber*) Displays on Air Synoptic

Packs may be in different modes

- If one pack standby and other normal
  - Pack in standby shuts down at "lower altitudes and higher ambient temperatures"
  - Occurs if ambient conditions do not allow standby cooling to be effective
  - Pack restarts in standby when conditions suitable for standby operation

If operating in standby mode on ground

- Pack is selected off until airborne
- When ambient air conditions allow for sufficient cooling, pack is selected on in standby mode
- Pack is shut down when descent to lower altitudes occurs

If both packs standby cooling mode OR one pack INOP and the other in standby cooling mode,

- Pack(s) operate continuously to maintain cabin pressurization
- No shutdown occurs at lower altitudes
- A single pack in standby cooling mode is sufficient to maintain pressure at cruise altitudes
- In standby cooling mode
  - Pack cooling capacity is reduced
  - Warmer temperatures in cabin and flight deck may result

<u>Note</u>:*Description of what happens in standby mode removed from OM; Previous OM editions described standby mode as heat exchanger only cooling*

Air Distribution

Flight deck air source

- 100% Comes from left pack only
- Flight deck pressure slightly higher than cabin
- Prevents cabin smoke from entering flight deck

Air Flow from Cabin and Flight Deck

- Air flows to lower deck
  - Then either exhausted through outflow valves ***or***
  - Drawn into lower recirculation system
- Air from lower recirculation fans
  - Mixes with pack air
  - Then enters lower deck distribution ducts
- Air from upper recirculation fans
  - Flows into lower deck distribution ducts

Recirculation Fans

- Either fan off causes packs to operate at full flow
- Cabin exchange rate is increased
- Fuel consumption increases <u>0.7% for each fan switch</u> turned **OFF**
- Recirculation fan switches—see next page

Temperature Control

Trim Air System

- Mixes hot trim air with cooled pack air to moderate flow to various zones
- Pack outflow determined by zone requesting coolest air

Trim Air Switches

- **ON**—Trim air valve is commanded open.
- Off (ON not visible)—Trim air valve is commanded closed, ***AND*** **FAULT** illuminates.
- **FAULT**
  - The trim air valve is failed closed ***OR***
  - The trim air valve is commanded closed because of a zone supply duct overheat ***OR***
  - Trim Air Switch is selected **OFF**.

Cabin Temperature Selector

- Sets master reference temperature between 65 and 85°F
- Flight attendants can vary individual zone temperatures by ±10°F from master temperature

Flight Deck Temperature Control

- Sets flight deck temperature in range 65-85°F (75°±10°)

If neither controller is available to pack temperature controller

- Pack outlet temperature is regulated to achieve either:
  - The <u>last temperature</u> set ***or***
  - An <u>average</u> cabin temperature of <u>75°F</u>

Complete loss of temperature controllers

- Can occur with:
  - Failure of both controllers or
  - Loss of electrical power
- Air supply and cabin pressurization controllers control pack flow control valves
- Pack flow rate is modulated to achieve pack outlet temperature in <u>40-110°F range</u>
- All flight deck temperature control disabled except pack control switches

Cargo Temperature Control

Forward Cargo Area

- Heated by warm air from forward equipment ventilation system
- No selectable temperature control available

Aft and bulk cargo areas—

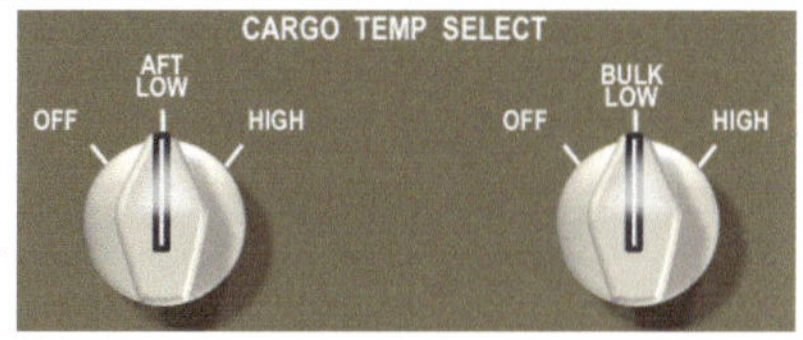

- Switches available for temperature regulation
- **OFF**—shuts off bleed air to compartment
- With temperature selector set to **LOW *and*** TAT less than 7°C (45°F)
  - Respective temperature control valve opens
  - Compartment temperature is maintained between 40 and 50°F.
- With temperature selector set to **HIGH *and*** TAT less than 21°C (70°F)
  - Respective temperature control valve opens
  - Compartment temperature is maintained between 65 and 75°F

Overheat Protection

- Automatically shuts down bulk cargo heating if overheat is detected
- If triggered, cargo heat to the related compartment cannot be restored in flight

Insulated curtain separates the two compartments

Equipment Cooling

Forward Equipment cooling

- Supplies cooling air to electrical and electronic equipment
  - On the flight deck
  - In the forward E&E compartment racks
- System uses internal fans and valves to
  - Draw cooling air from cabin and force it through equipment racks and flight deck equipment
- Vents warmed air to two places
  - Directly out outflow valve ***OR***
  - Directs some of the air into forward cargo* compartment if it requires additional heat
- If system is inoperative on the ground
  - Ground crew call horn sounds
  - EICAS advisory message **EQUIP COOLING**

Override Mode

In flight—forward system is **automatically** reconfigured when ***<u>any</u>*** of the following occur (***SEFFF—Smoke, Equipment, Flow, Fans, Fire***:

- ♦ ***<u>Smoke</u>*** is detected in the forward equipment cooling system or forward equipment ventilation system
- ♦ <u>***Equipment*** cooling</u> switch is **OFF**
- ♦ <u>Low ***Flow***</u> is detected in flight
- ♦ <u>Both supply ***Fans*** fail</u> in the forward equipment ventilation system in flight
- ♦ <u>Forward cargo ***Fire***</u> extinguishing system is **ARMED**

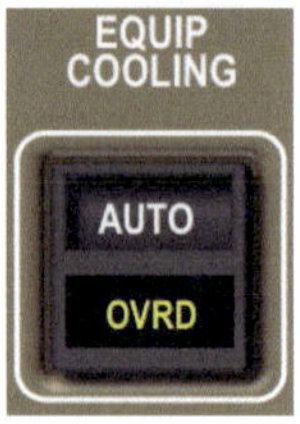

Manually commanded by the equipment cooling switch when placed in **OVRD**
- In flight, use of this switch "aids smoke evacuation from flight deck"

System changes in override mode
- Both supply fans off
- Vent valve opens
- Cabin differential pressure is used to reverse airflow and exhaust air (and smoke, if present)
- Air exits through override valve to an overboard vent

AUTO—system operating normally
- **AUTO** off (not in view)—any of the following exist:
  - ♦ Both equipment cooling supply fans are ***not*** operating
  - ♦ Override valve is open
  - ♦ Forward cargo heat valve is closed
  - ♦ Airflow in Override mode

♦ **300ER** Nitrogen generation system (NGS) is shut down

OVRD—on when ECS configures to override mode:
- ♦ Both equipment cooling supply fans are failed, ***OR***
- ♦ Smoke is detected in the equipment cooling system ***OR***
- ♦ **EQUIP COOLING** switch is off ***OR***
- ♦ **CARGO FIRE ARM FWD** switch is in **ARMED**
- Flow is "adequate" in cruise
- Decreases as airplane descends due to lower differential cabin pressure

Aft Equipment Cooling
- System provides
  - Cooling and ventilating air to aft electronic equipment
  - Ventilating air for lavatories and galleys
- Two fans—primary and backup
- Conditioned air is supplied to galleys from air distribution system
- Aft equipment cooling system fans draw air from
  - ♦ Aft electronic equipment ♦Galleys ♦Lavatories
- Air discharges through aft outflow valve

Recirc Fan Switches

- Two switches—
  - Upper Switch controls forward and aft upper Recirc fans
  - Lower Switch controls left and right lower Recirc fans
- When one or both recirculation fan switches are **OFF**
  - Packs operate at full flow
  - Cabin air exchange rate is increased
  - Fuel consumption increases 0.7% for each fan switch **OFF**.

Pressurization

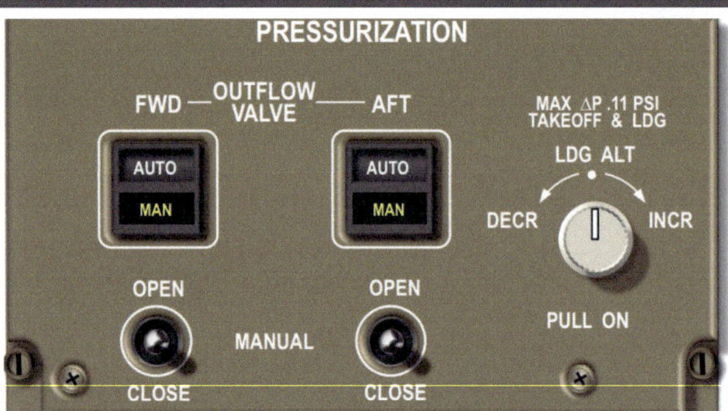

Outflow and Control Valves

Two outflow valves
- Forward and aft
- Most airflow is out aft valve—improves ventilation and smoke removal
- Either valve can control pressure if other fails

Two positive pressure relief valves

Four negative pressure relief doors

Automatic operation

Selected when outflow valve switches are set to **AUTO**

Inputs from
- ♦ Ambient pressure on ground
- ♦ Flight plan data from flight management computer
- Uses these to calculate pressurization schedule

Pressurization Schedule
- Small positive pressure on ground
- Increases to maximum cabin differential pressure
- Climb of cabin then becomes variable with airplane climb
- At maximum certified cruise altitude, cabin pressure is 8,000 feet pressure altitude
- On descent, pressure is gradually increased to slightly below landing field pressure altitude
- On touchdown, both outflow valves open to depressurize cabin

Landing Altitude
- Departure field altitude is retained in memory in case of return to point of origin
- Arrival field is planned descent point after:
  - ♦ 400 NM from departure airport ***OR***
  - ♦ Half way to destination
  - *Whichever occurs first*

LANDING ALT EICAS Message—Indicates
- ♦ Landing altitude is unavailable from the FMC
- ♦ No landing altitude set in MAN
- Cabin altitude controller assumes a landing altitude of 2000 feet MSL

Manual Operation

Selected when outflow valves are set to **MAN**

Outflow valves then manually operated by moving related valve switch to open or close

Valve position displays on EICAS when in manual mode
- Moving from full open to full closed takes 30 seconds

Landing altitude selection
- Can be accomplished with landing altitude selector
- Pulling switch out
  - Removes FMC landing altitude
  - Displays pressurization indications on EICAS

Operation with Loss of Cabin Pressure

If Sudden loss of pressure occurs
- Outflow valves close to control pressure loss
- When inside and outside pressures equalize, outflow valves open
- Protects airplane from negative overpressure

Crew should not manually close valves in this case

Pressurization System Indications

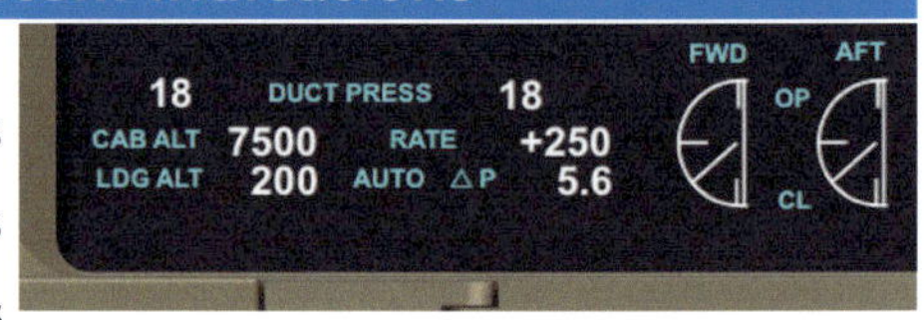

Appears on EICAS for abnormal pressurization events

Displays when (associate with items on display)
- ♦ Duct pressure is below normal range (*Amber*) with respective engine running
- ♦ Cabin altitude is above normal range (*off schedule*)
- ♦ EICAS **CABIN ALTITUDE** message is displayed (*Red*: >10,000')
- ♦ Landing altitude selector is pulled **ON**
- ♦ Cabin differential pressure above normal (*Amber*)
- ♦ Cabin differential pressure is excessive (*Red*)
- ♦ Air synoptic is on any MFD
- ♦ An outflow valve switch is selected to **MAN**
- ♦ Abnormal air-related EICAS messages have occurred:
  - **CABIN ALTITUDE AUTO**
  - **LANDING ALTITUDE**
  - **OUTFLOW VALVE FWD**
  - **OUTFLOW VALVE AFT**

Key Cabin Pressure Altitudes

- **>8500' Cabin pressure**—Display turns amber
- **>10,000' Cabin pressure**—Display turns red
- **>13,500' Cabin pressure**—PSU Masks drop automatically

Miscellaneous

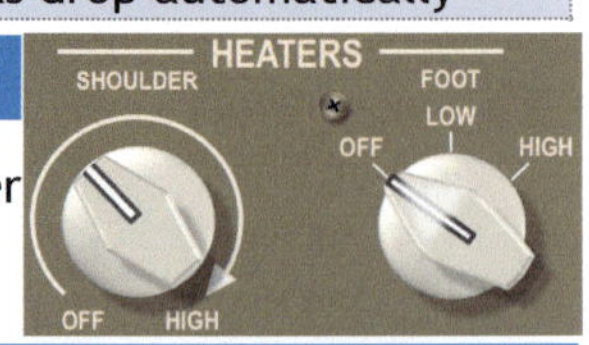

Shoulder and foot heaters

Provided for Captain and First Officer seats

Acronyms

ASCPC ....Air supply cabin pressure controller
PRSOV ....Pressure regulating and shut off valve

Anti-Ice and Rain Protection

Automatic Ice Detection System

General Features

Detects icing in flight
Inhibited on the ground
Signals sent to both engine and wing anti-ice ***when:*** In flight ***and***
Systems are in **AUTO**
Two probes
One on each **side of forward fuselage**
Automatically deiced **if ice-buildup is sensed**

Operation

When either probe detects ice buildup more than once
Engine anti-ice valves automatically open
Engine cowls are de-iced
When either probe detects ice buildup several times
Wing anti-ice valves automatically open
Six wing leading edge slats (three per wing) de-iced
In both cases, valves close when system no longer detects icing

Engine & Wing Anti-Ice Systems

Engine Anti-Ice System

Bleed air heats engine cowls only when ice is detected
May be operated in flight or on the ground
In flight, AUTO normally used
On ground, manual (**ON**) must be used
Each engine has its own independent system
Leak detection
Dual loop anti-ice duct leak detection system on each engine
If duct leak is detected, affected engine anti-ice valve closes

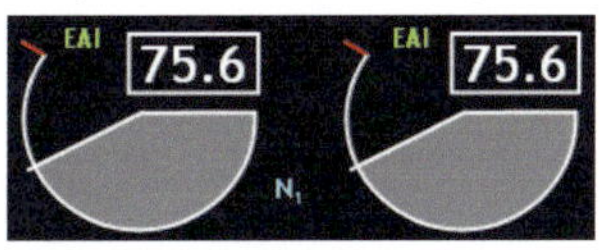

Engine anti ice annunciation is on N_1 gauge of EICAS
EAI Indication shows engine anti-ice ON

Wing Anti-Ice System (WAI)

Can be operated in flight only
Inhibited on ground, regardless of switch position
Automatic Wing Anti-Ice Inhibits—when ALL of the following:
♦ TAKEOFF Mode is selected ***AND***
♦ Less than 10 minutes has elapsed from liftoff
Wing Anti Ice Inhibits
♦ If TAT > 10°C, WAI is inhibited for 5 minutes after liftoff
♦ **TAKEOFF** mode is selected, ***and*** less than 10 minutes has elapsed after liftoff
Mid-Wing Leading Edge Slats deiced by WAI system
3 LE Slats (out of 7) on each wing
Slats are numbered from left wingtip across to right wing tip

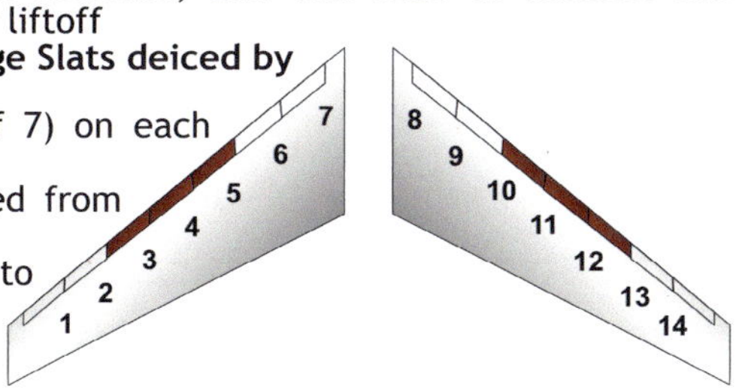

Leading edge slats
3, 4, 5, 10, 11, & 12 are deiced (*Slat numbering from previous OM*)
Symmetry Protection
If a bleed source is lost ***and*** duct isolation has ***not*** occurred:
Isolation valves open automatically to maintain anti-icing to both wings
If one wing anti-ice valve fails closed:
Other is automatically closed to prevent asymmetric wing anti-icing

Leak Detection Systems
System senses and isolates duct leaks automatically
See *AIR section* for details
Wing Anti-Ice Annunciation (WAI)
Displayed below EICAS N_1 indications when a wing anti-ice valve is open

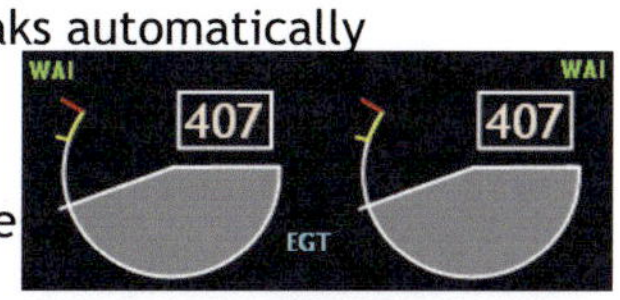

Window Heat

Forward Windows

Exterior surface anti-icing
Interior surface anti-fogging
Backup antifogging system operates automatically in the event of a primary system failure

Side Windows

Interior surface anti-fogging only

Window Heat Switches (4)

Control heating for all flight deck windows
With switches ON, window heat operates as soon as power is available
Thermal shock prevented by gradual increase in heating when power is applied

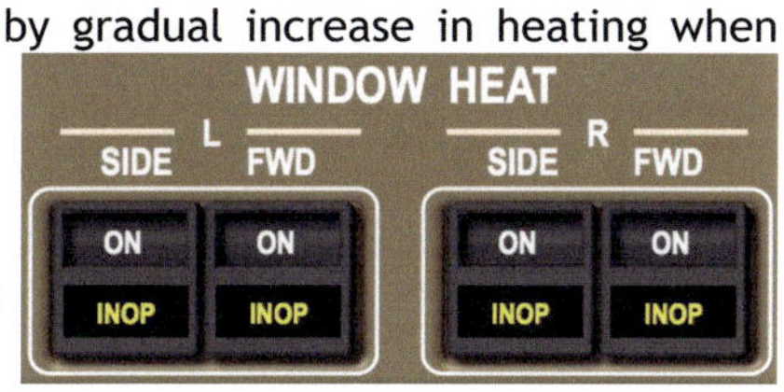

Switch positions
Pressed—ON;
Window heat is applied to selected windows
This is normal position
Released—Off, **INOP** indication illuminated
INOP—(*Amber*) Switch is off ***OR***
Overheat is detected ***OR***
System fault has occurred

Backup Window Heat Switches
Overhead Maintenance Panel
Allows removal of power in the event of window damage
Controls arcing and smoke danger
Documented—QRH **Window Damage L, R** procedure ....See QRH AIRPLANE GENERAL 1.21

Windshield Wipers

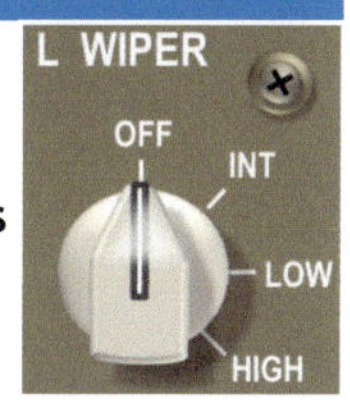

L and R Switches
Self Explanatory

Probe Heat

Components heated when either engine is running
3 Pitot Probes
2 Angle of Attack vanes
Component heated only in flight—TAT probe

Acronyms

WAI ....Wing anti-Ice
EAI ....Engine anti-ice
PRSOV ....Pressure regulating and shut off valve

Autoflight

Editor's Note: *This section covers systems-related notes which are considered helpful for understanding the relationship between the autopilot and other aircraft systems. No attempt is made to summarize the operation of the autoflight system, as it is better learned in the CPT and through Plato than through this medium. In addition, normal interaction with the system keeps pilots aware of the capabilities, techniques and options which work well in flight. For that reason, notes here are intended for systems review more than for initial instruction.*

Overview

Components of Automatic Flight Control System

Autopilot Flight Director System (AFDS)
Autothrottle System

Control

Mode Control Panel (MCP)
- Select mode engagement
- Insert clearance specifics such as heading and altitude
- Override computer planned parameters
- <u>Selected</u> altitude (*as well as actual aircraft altitude*) is transmitted to ATC

Flight Management Computer (FMC)
- Upload and activate route for flight plan
- Set planned parameters, such as
  - Crossing altitudes
  - Cruise speed and altitude
  - Performance and weight and balance parameters
  - Takeoff data

Autopilot

Autopilot Flight Director System

Components
- 3 Autopilot flight director computers (AFDCs),
  - Control flight directors & autopilots
  - Does not control flight control surfaces directly
  - Sends inputs to the autopilots,
  - Autopilots send commands to flight control surfaces through fly-by-wire flight control system
  - Rudder control
    - ***Not*** normally accomplished through autopilot ***except***, during
      - Autopilot approach and landing
      - Nosewheel steering and rudder during rollout
    - For more information on rudder control, see Flight Control chapter, *page 41* of this study guide
- Mode Control Panel (MCP) which controls
  - Autopilot
  - Flight director
  - Altitude alert
  - Autothrottles
  - Select and activate AFDS modes
  - Setting of
    - Altitudes
    - Speeds
    - Climb and descent profiles

Mode Selection

- Press mode button
  - Green bar in bottom half of mode switch illuminates
  - Confirms mode armed or engaged
- Pressing any other button deselects original mode and selects new one

Exception:
- **APP** mode will not deselect after capturing localizer and glide slope
- To disengage **APP**:
  - Disconnect autopilot ***and*** turn off both flight directors
  - ***OR*** Engage go-around mode
  - ***OR*** if above 1500' radio altitude, press APP mode switch again
    - Default roll and pitch modes result

Some mode switches can be disengaged by pressing a second time:
- ♦ **VNAV** ♦ **LNAV**
- ♦ **LOC** ♦ **APP** (*See notes above*)

Autopilot

Engagement—Press either of two MCP autopilot engage switches

Disengagement
- Four ways to disengage
  - ♦Press either control wheel autopilot disengage switch
  - ♦MCP autopilot disengage bar
  - ♦Override with control column or control wheel
  - ♦With LAND2 or LAND3 annunciated, can override with rudder pedals as well

Results
- EICAS warning, **AUTOPILOT DISC**
- Warning results whether manual or automatic disconnect occurs

Degradation
- Autopilot detects mode degradation
- Remains in attitude stabilizing mode based on inertial data
- Persistent degradation results in
  - Amber line through modes on affected flight mode on PFD
  - EICAS Caution **AUTOPILOT** message
- If degradation clears, mode annunciation displays inside green box

Autopilot engagement is recommended with less than 4000 RVR (1200 m) or 3/4 mile visibility

Uncommanded Turns in LNAV

Have been reported

When observed, FMC CDU prematurely sequenced a point
- May occur when within ***1 NM*** of active waypoint for small route modifications, ***OR***
- Within ***4 NM*** of active waypoint for modifications involving a large turn

Prevention—Avoid:
- ♦Executing lateral OFFSET when approaching an active waypoint
- ♦Entry of a vertical or lateral flight planning change when approaching an active waypoint.
- ♦Executing DIR-TO with ABEAM selected when approaching an active waypoint.

Flight Director

Flight Director Display

Steering bars in view when
- Respective (L/R) FD switch is on
- If switch off, pressing a TO/GA switch displays FD bars

Degradation
- If degradation sensed, FD provides attitude stabilizing commands based on inertial data
- Persistent degradation results in removal of steering bars from PFD
- Mode self clears if fault disappears

Stall or Overspeed Protection also result in FD bars disappearing from PFDs

AFDS Modes

Status Annunciation—appears on PFD above attitude indication
- **FLT DIR**—FD is on and A/P ***is not*** engaged
- **A/P**—FD is on and A/P ***is*** engaged
- **LAND3**—3 A/Ps are operating normally for autoland
  - A single fault cannot prevent the autopilot system from making an automatic landing
  - Status is considered fail operational
- **▶ LAND2 ◀** —AFDS redundancy is reduced
  - In some cases, only 2 A/Ps are available
  - A single fault cannot cause the autopilot system to make significant deviation from flight path
  - Status is considered fail passive
- **NO AUTOLAND**—AFDS is unable to make autolanding
- Changes in status
  - Highlighted with aural alert and
    - Amber box around **NO AUTOLAND** for 10 seconds
    - Green box around **▶ LAND2 ◀** for 5 seconds
  - Alerts inhibited until after touchdown if non-safety of flight
  - Below 200 feet RA, all except **NO AUTOLAND** are inhibited

Mode Annunciation—appears just above PFD AFDS status annunciation
- Three sections, left to right, are

Autothrottle	Roll	Pitch
SPD	LNAV	VNAV
	LOC	G/S

  - Mode Indications
    - Green large letters at top—Engaged
    - White smaller letters at bottom of block—armed

Autothrottle modes
- **THR**—Thrust applied to maintain V/S required by engaged pitch mode
- **THR REF**—Thrust set to limit displayed on EICAS
- **IDLE**—Displayed while autothrottle moves to idle
  - When at idle, mode changes to **HOLD**
- **HOLD**—Autothrottle servos disabled
- **SPD**—Autothrottles adjust to maintain set speed
- Note: *A/T selection inhibited on takeoff from 50 KIAS to 400'AGL*

Roll Modes
- **LNAV** Armed—Will engage when parameters met
  - Engaged—LNAV is tracking active route displayed on the ND if aircraft is:
    - Above 50 feet AGL (radio altitude) and
    - Within 2½ NM of active route leg
- **HDG**
  - **HDG SEL**—Airplane is turning to or tracking heading set in MCP heading/track window
  - **HDG HOLD**—If wings level when **HOLD** selected, AFDS holds that preset heading
    - If in a turn when **HOLD** selected, holds heading at rollout
- **TRK**, **TRK SEL**, **TRK HOLD**—Same as **HDG** above, except maintaining the track, not the heading
- **LOC**
  - Armed—AFDS captures the localizer when
    - Within range ***and***
    - Within 120° of localizer track
  - Engaged—AFDS maintains localizer course
- **TO/GA**
  - Ground—Mode is engaged by selecting either FD **ON** when both FDs are **OFF**
    - Mode guidance becomes active at liftoff
  - Flight—Armed with flaps not retracted OR at glideslope capture
    - No annunciation for armed in flight
    - Engaged by pressing a **TO/GA** switch
    - AFDS maintains ground track present at mode engagement
- **ROLLOUT**
  - Armed—Displayed below 1500' RA
  - Engaged—Engagement occurs at 2' RA
    - Rudder and NW steering used to maintain aircraft on centerline

Pitch Modes
- **TO/GA**
  - Ground—Engaged when
    - Either FD is turned on when both had been off ***OR***
    - By pressing either **TO/GA** switch with airspeed over 80 knots
    - Roll guidance active at lift-off
    - **Note:** *Pressing* **TO/GA** *> 80 knots IAS disarms* **LNAV** *and* **VNAV**
  - After takeoff, AFDS commands pitch to maintain
    - V_2 (based on speed set before takeoff) plus 15 knots ***OR***
    - If airspeed remains above target for 5 seconds, target reset to current speed ***OR***
    - IAS/Mach window speed if the value is changed to a speed greater than the target speed
  - In flight other than takeoff, armed when flaps not retracted (no annunciation in flight)
  - After go-around initiated
    - Commanded speed is the higher of
      - MCP IAS / MACH window ***OR***
      - Current speed ***to a maximum of*** MCP IAS / MACH window speed plus 25
    - GA is displayed as thrust limit on primary EICAS engine display
    - **Note: 300ER** *During go-around from* **LAND 2** *or* **LAND 3** *approach, automatic LNAV activation causes disengagement of autopilot rudder control. If executing an engine out missed approach with thrust asymmetry compensation inoperative, manual rudder control may be required to prevent large roll and yaw excursions.*
- **VNAV**—Armed by pressing **VNAV** switch
  - Engages above 400' on takeoff if armed
    - **Note**: *400' is referenced to baro altitude recorded at 100 knots on takeoff roll.*
  - **VNAV SPD**—Maintains FMC speed displayed on PFD and/or CDU **CLIMB** or **DESCENT** pages
    - If speed intervention used, MCP **IAS/MACH** selector is used to set speed
  - **VNAV PTH**—AFDS maintains FMC altitude or descent path with pitch commands
    - As aircraft approaches top of descent with cruise altitude still set, CDU scratchpad message **RESET MCP ALT** displays
  - **VNAV ALT**—Altitude is maintained or captured
    - If conflict occurs between VNAV profile and MCP altitude
      - Airplane levels at MCP altitude ***and***
      - Flight mode annunciation becomes **VNAV ALT**
- **V/S** Pressing mode switch
  - Opens vertical speed window to display current vertical speed
  - Allows changing commanded vertical speed with **V/S** wheel
    - Pitch maintains rate of climb or descent selected in **VS/FPA** window
- **FPA**—Selected by pressing **VS/FPA** mode button ***and***
  - Pressing small black **VS/FPA** selector button on MCP
  - AFDS maintains pitch as required to track commanded flight path angle
- **FLCH SPD**—Selected by pressing mode button
  - IAS/MACH window opens and displays current IAS
  - AFDS adjust pitch as needed to maintain selected IAS/MACH window airspeed or mach
- **ALT**—Altitude hold mode is engaged
  - Engaged by pressing MCP altitude hold switch, ***OR***
  - Capturing the selected altitude from a **V/S**, **FPA** or **FLCH** climb or descent

G/S—AFDS is tracking ILS glideslope
Can capture when within 80° of localizer course
Capture may occur before localizer capture

FLARE
Armed—on autoland, displayed below 1500' RA
Engaged—between 60 and 40 feet RA
Accomplishes autoland flare maneuver so the AFDS can transition to **ROLLOUT** mode

Autothrottle System

Autothrottle operation

Thrust control from takeoff through landing
Can be used without FD or Autopilot
Available modes include
THR REF **SPD** **HOLD** **IDLE**
Use during manual landing
Thrust reduces to idle at 25' RA
This occurs in any mode ***except*** **TO/GA** or **FLCH**
Can be overridden at any time by pressing either A/T disconnect switch on thrust levers

Autothrottle Automatic Arming—occurs automatically if:
All of the following conditions are met:
- Autothrottle is functional (no system faults)
- Airplane altitude above 100' RA on approach or airplane barometric altitude 400 feet above airport on takeoff
- No autopilot or flight director is active OR
An A/P or FD is in **VNAV**, **FPA**, **ALT**, **V/S**, or **GS**

And both of the following occur:
- Speed less than FMC calculated for one second
- Thrust below reference thrust

Automatic Disarming—occurs automatically if:
Fault in active A/T mode detected
Override—Thrust Levers overridden manually after A/T has begun to retard to idle
Reverse lever—either one raised to reverse idle
Shut Off—engines both shut down

Autothrottle "Wake Up" Mode
Allows stall protection when armed but not activated
Activates when speed decreases to near stick shaker activation
Advances in **SPD** or **THR REF** modes to maintain the greater of
- Minimum maneuvering speed (top of amber band) ***or***
- Speed set in the mode control speed window

Does ***not*** activate in the following conditions
- Pitch modes **FLCH** or **TOGA** ***OR***
- Airplane below 400' AGL (takeoff) or below 100' AGL on approach

Note: *During descent in* **VNAV SPD**, *autothrottle may activate in HOLD mode and will not support stall protection*

Abnormal Operation

EICAS Messages for servo failure
AUTOTHROTTLE L or **R** —respective servo failure
PFD Status message for one A/T only armed
L or **R** precedes status, indicating only that A/T is engaged in the specified mode
Example: **L SPD**

Autothrottle Uncommanded Motion

Has been reported during cruise flight
Prevention
Prior to engine start, move the FMC selector switch to either L or R, then back to AUTO
Cycling of the switch inflight at any time will resolve or prevent the anomaly

Automation Key Altitudes & Speeds

Takeoff

Below 50 Knots—Pressing **TO/GA** switch engages thrust reference (**THR REF**) mode
50 knots or above—If A/T not engaged, cannot be engaged until above 400'
80 Knots—Mode changes to **HOLD**
Over 80 Knots—Pressing a **TO/GA** switch disarms **LNAV** and **VNAV**
100 knots—FMC records barometric altitude; used for:
Engaging **VNAV** if armed when 400' above this altitude
Enabling autothrottle activation if not active
Commanding acceleration altitude for flap retraction
Set climb thrust if altitude entered on MCP Takeoff page 2
If engine failure occurs on the ground, pitch command target speed at liftoff is the greater of V_2 ***OR***
Airspeed at liftoff

Climb

Lift Off—Pitch commanded to maintain:
Target speed of V_2+15
If speed is above target speed for 5 seconds, target is reset to current airspeed limited to V_2+25
If engine failure occurs after lift-off, pitch command target speed is:
If airspeed is below V_2........ V_2
If airspeed between V_2 and V_2+15........Existing speed
If airspeed above V_2+15........ V_2+15
After Liftoff—if **TO/GA** pressed with A/T engaged:
Thrust Lever derates are removed
A/T engages in **THR REF**
50' RA—**LNAV** engages if armed
400 feet above active runway elevation
VNAV engages if armed
A/T sets selected reference thrust and annunciates **THR REF**
Acceleration height—pitch commands speed to takeoff flap placard speed minus 5 knots
Flap retraction completion—**VNAV** commanded speed is greater of:
250 Knots or
V_{REF} + 80 Knots
Thrust reduction point
Altitude in MCP takeoff page 2 ***OR***
Flap retraction
FMC changes thrust limit to armed climb limit—**CLB**, **CLB1**, or **CLB2**

Communications

Cockpit Voice Recorder System

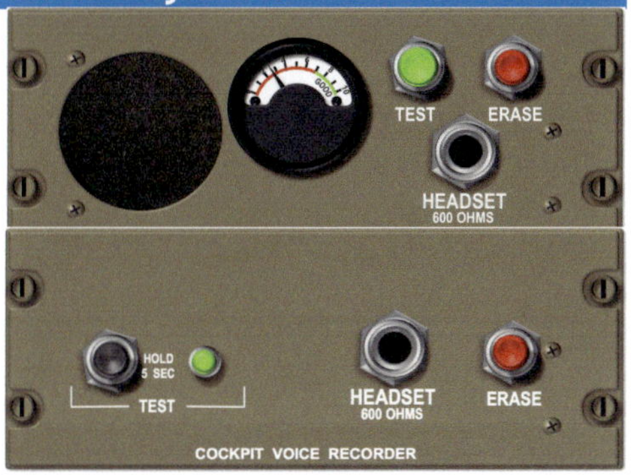

Records inputs from:
- Flight deck through area mic
- Communications channels from any audio control panel

300ER Status Indicator in place of meter—illuminates after successful test
- Extinguishes after test button is released

Records continuously
- **300ER** Also records datalink communications
- **300ER** Records for 10 minutes after AC power is off

Cockpit Voice Recorder Circuit Breaker
- Location—E&E P110 panel (F6—labeled **VOX RCDR**)
- Procedures requiring pulling breaker
  ....AOM 16.7 Cockpit Voice Recorder (CVR)

VHF Communications

Three receiver/transmitters
- Left—Configured for voice only
- Center—Normally configured for data for ACARS
- Right—Normally voice, may be configured for data,

To change from DATA to voice
- Select a voice frequency
- Transfer frequency to active side of window

200 Hot Mic Protection
- Ground with both engines shut down
  - A VHF which transmits more than 35 seconds
    - Automatically disabled
    - Dashes in the tuning panel window for that radio
  - Radio is reactivated when transmit button for that radio is released
- Any other time—if VHF or HF radio transmits over 30 seconds
  - EICAS message—**RADIO TRANSMIT**
  - Message disappears after transmission ceases

300ER Hot Mic Protection—If a VHF radio transmits > 35 seconds:
- ♦ That radio is automatically disabled AND
- ♦ Dashes appear in the tuning panel frequency window AND
- ♦ Intermittent tone is heard through respective radio audio

HF Radios

Two radios, L and R

One antenna
- May be used for reception on both simultaneously
- Only one HF may transmit at a time—other deactivated during transmission

Tuning
- First time transmitter is keyed after new frequency entered, frequency is tuned
  - Tone is heard while tuning takes place
  - Maximum tuning time is 7 seconds
  - A tone lasting over 7 seconds indicates a failure to tune
- Last 100 frequencies are stored and tune quickly
- If a frequency not in last 100 is tuned
  - Longer tone as frequency is tuned
  - Frequency is added to most recent 100 list

Communications Crew Alerting System

Provides visual and aural alerts to call crew's attention to various issues

Visual alerts—EICAS message with bullet symbol, i.e., ▣ COMM

Aural Alert—High-low chime

Three Levels of Alert
- High—not implemented currently
- Medium—Visual and aural alert
  - Crew awareness required
  - Crew action may be required
- Low—Visual alert only

SELCAL System (Selective Call)

Monitors 3 VHF and 2 HF radios

Alerts crew through communications crew alerting system

ACARS

All functions accomplished through ACARS Manager
- When appropriate, use buttons on MFD screens ***or***
- Interactive buttons on glare shield
  - See COMMUNICATIONS 10.14, and 50.1 – 60.3

SATCOM (Satellite Communications)

Allows both voice and data communications

Three channels
- Two dedicated for voice transmissions
- One dedicated for data (ACARS)

EICAS Message ▣ SATCOM MESSAGE
- Indicates a SATCOM message requiring their attention on one of the CDU SATCOM pages
- Servicing message clears EICAS message

Worldwide SATCOM System
- Airplane Earth Station (AES) Segment
  - Airplane component
  - Includes antenna continuously aimed at satellite
- Ground Earth Station (GES) segment
  - Ground component
  - Initiates and receives through ground antenna aimed at satellite
- Satellite Segment—relays signals between AES and GES

Interphone Communication System

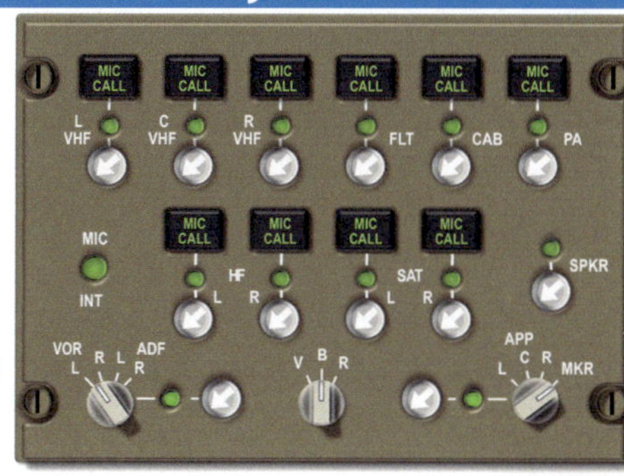

Flight Interphone System
- Allows communication:
  - Between flight deck occupants
  - Between flight deck and ground crew
- Selected with **INT** position of audio control panel
  - **CALL** illumination in **INT** button
    ....indicates ground crew call
  - • **GROUND CALL** EICAS also indicates ground call

Service Interphone System
- Voice communications allowed between ground crew stations at various locations around airplane
- Connected to flight interphone system with Service Interphone Switch on overhead panel

Cabin Interphone

Some of the following notes are from an unofficial handout, but the data has been confirmed to work

Normal Calls—
- ♦ Single hi-lo chime ♦ EICAS msg: CABIN CALL

Door 1L/1R	11	Single Door, L	7<Door #>
Door 2L/2R	22	i.e. Door 1R	71
Door 3L/3R	33	Single Door, R	8<Door #>
Door 4L/4R	44	i.e. Door 4R	84

Priority Calls
- ♦ Single hi-lo chime ♦ EICAS msg: ▣ CABIN ALERT

Conference Calls

Multiple Calls, or clearing Calls when out of sync with incoming calls:
- To delete multiple calls in queue use Center CDU
- Press **DELETE**, then line select 1R or line showing the call to be deleted

Electrical Systems

System Overview

Main Electrical Sub-Systems

Main AC power—Primary source of all airplane electrical power
Backup Power
DC Power
Standby Power
Flight Control Power

Automated Features

System operation is automatic
Faults are automatically detected and isolated

AC Electrical System

Electrical Load Management System (ELMS)

Provides load management and protection
Ensures power is available to critical and essential equipment
Load Shedding
- If loads exceed power available, sheds AC loads
- Uses priorities to download until loads are within capacity available (airplane and/or external)
- Order of shedding:
  - ♦ Galleys
  - ♦ Utility busses
  - ♦ Individual equipment items powered by main AC busses
- When additional power available or loads decrease, shed loads restored to power in reverse order
- Synoptic Message—**LOAD SHED** Displays when conditions exist

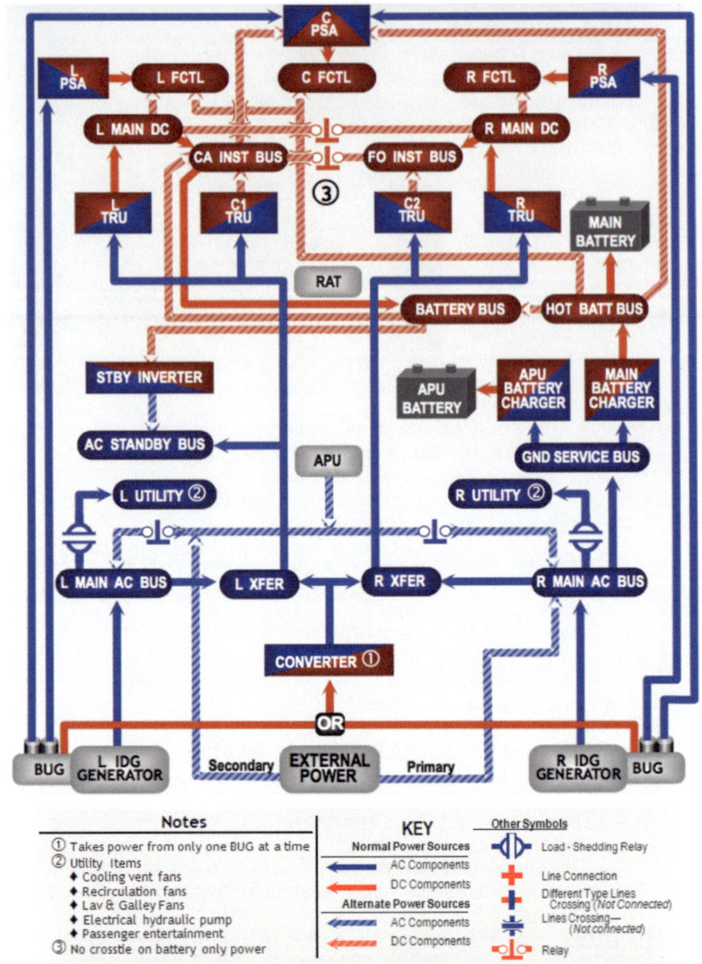

AC Electrical System Power Sources

Aircraft Sources
- L & R Engine integrated drive generators (IDG)
- APU Generator

External—Primary and secondary external power
Requirements—Entire AC load can be supplied by any two main AC sources
Parallel versus Independent Power
- Power sources are normally isolated from each other
- During power source transfers on the ground
  - Operating sources momentarily paralleled
  - Prevents power interruption

Integrated Drive Generators (IDGs)

One on each engine—Automatic control and system protection functions
Connection to AC Busses—On engine start:
- IDG automatically powers respective main bus
- Previous power source is disconnected

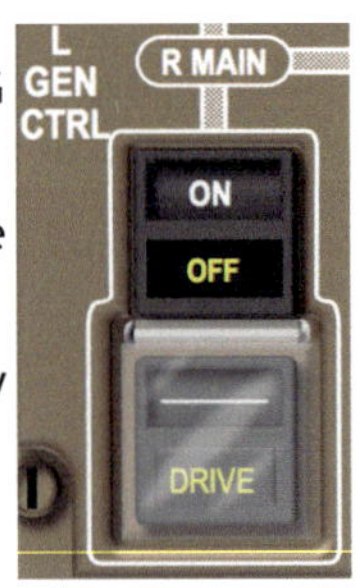

Disconnection from AC Busses—An operating IDG may be disconnected from its bus by:
- Pressing the generator control switch to **OFF**
- Selecting an available external power source prior to engine shutdown

Generator Control Switch
- **ON**—Arms generator breaker to automatically close when generator power is available
- **OFF**—Opens field and generator breakers
  - Resets trip fault circuitry

Generator Drive Light

DRIVE—
- *Without* generator **OFF** light
  - Indicates low oil pressure on respective IDG
  - EICAS message **ELEC GEN DRIVE L** *or* **R** displays
- *With* generator **OFF** light
  - Indicates high oil temperature on respective IDG
  - Disconnection is automatic in this case
- ***Pressed***—Disconnects IDG Drive from Engine
  - IDG Cannot be reconnected by flight crew

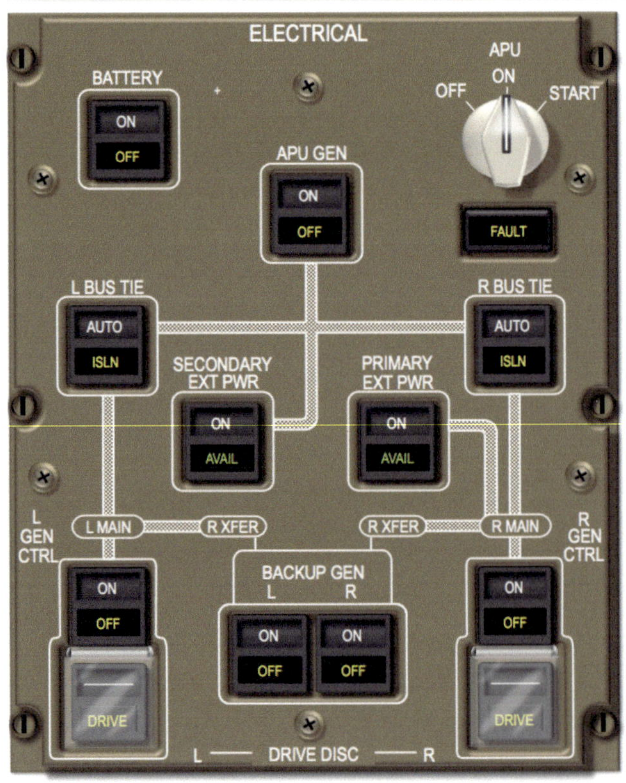

Electrical Synoptic

Comes up when ELEC synoptic display button is pressed

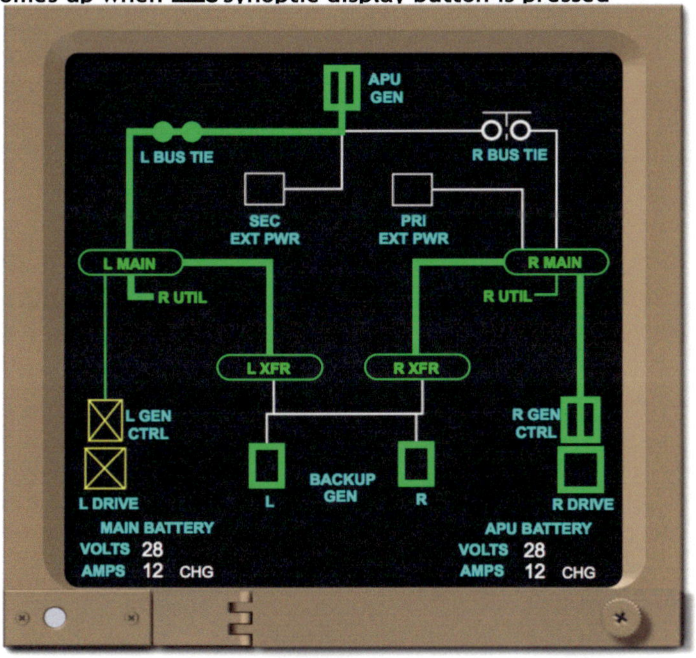

Electrical Synoptic Key

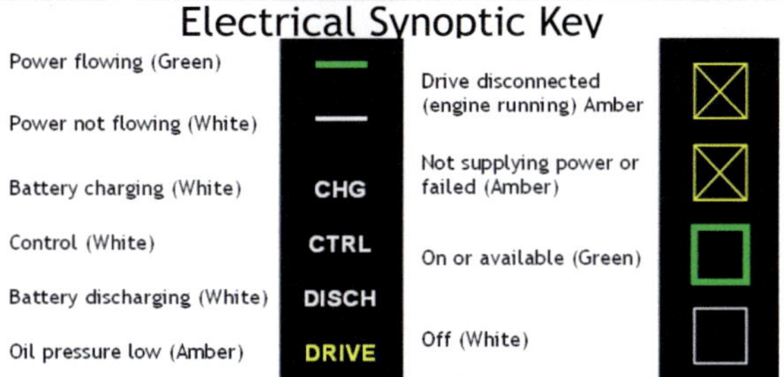

APU Generator

Electrically identical to IDG generators
Can power either or both main busses
May be used in flight as a replacement for an IDG source
APU Selection Priority
- If no other power source is available, APU generator automatically connects to both main AC busses
- If primary external is powering both main busses,
  - APU powers left main AC bus
  - Primary external source powers right main bus
- If primary external is powering the right main bus ***AND***
  - Secondary external is powering the left main bus, When APU comes on line it replaces the secondary and powers the left main bus
- If secondary external is powering both main busses when the APU comes on line, the APU will replace it and power both main busses

APU Generator Off Light indicates
- APU generator is operating, but the generator breaker is open due to either a fault ***OR***
- APU Generator switch is selected OFF

Bus Tie Breaker Priorities

- ♦ Respective side
- ♦ APU
- ♦ Opposite Side

External Power

Two types of external connection
- Primary
- Secondary
- Either is capable of powering both main AC busses
- Some items will be load-shed due to capacity restrictions

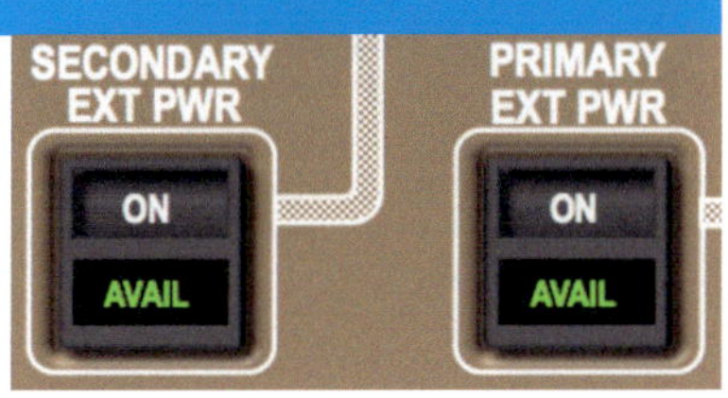

Connecting power
- Different for primary and secondary
- Primary—**AVAIL** light will illuminate when primary external power is connected, regardless of other switch positions
- Secondary—**AVAIL** light only on when battery switch is **ON** ***OR*** primary power is available

AC Bus Power Distribution

Source/Bus	Normally Powers
Right IDG	**R Main AC Bus**
Left IDG	**L Main AC Bus**
Primary External (when connected)	**R Main AC Bus**
Secondary External (when connected)	**L Main AC Bus**
APU	Left or Both Unpowered Main **AC Busses**
R Main AC Bus	**Ground Service Bus**
APU Generator ***OR*** Primary External Power	**Ground Handling Bus** (*No other sources can power this bus*)
Main AC Bus (Each)	Respective **Transfer Bus** Respective **Utility Bus** Respective **Galley Bus**
Transfer Busses	**DC Transformer-Rectifiers** **AC Standby Bus**
Utility Busses	Forward Galley Heater, Chiller Boost fan Also: CA & FO Area Heaters, Lav Heaters & Shaver Outlets, Gasper Fan
Ground Service Bus	**Main & APU Battery Chargers** Also: Miscellaneous cabin and system loads, cabin lighting and outlets, Left forward fuel pump

Isolation for Autolandings occurs when:
- Approach mode is engaged and
- Aircraft passes through 1500' radio altitude

Auto-Pilot	Powered By
Captain	Left IDG through the L AC Transfer Bus and the L Main DC Bus and the Captain's Flight Instrument Bus
Center	Right IDG through the Main Battery Charger and the Battery Bus and the AC Standby Bus
First Officer	Backup System, which powers R AC Transfer Bus and the R Main DC Bus and the First Officer's Flight Instrument Bus

Electrical System Autoland Bus Isolation

(Selected Major Components)

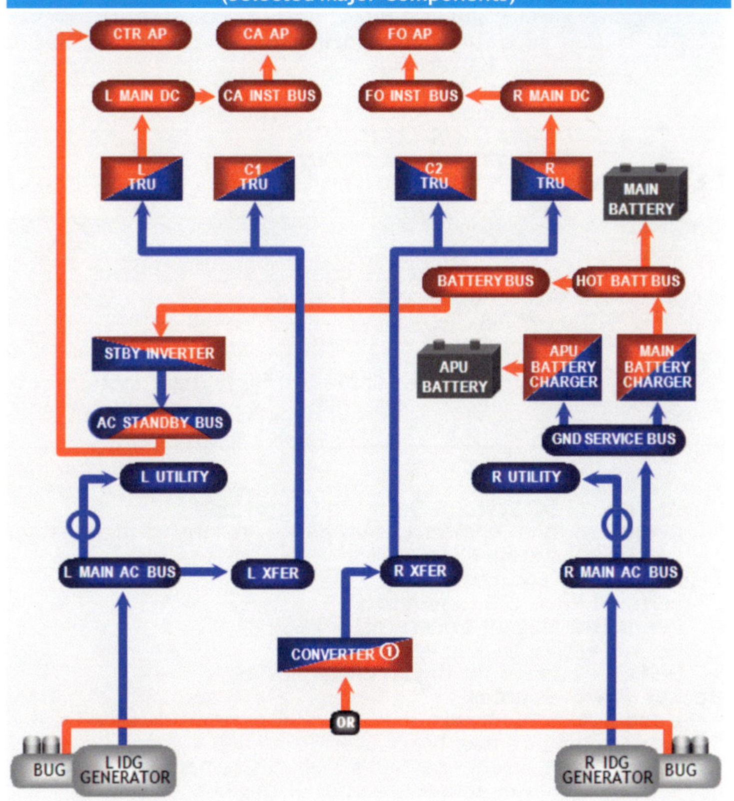

Backup AC Electrical System

Provides Power to Key Sub-Systems

Automatically activated when:

Only one main AC Generator available (including APU)
Power to one or both of the main AC busses is lost
Approach (**APP**) is selected for autoland
System is tested after engine starts

System transfers power without interruption

Components

Backup Generators (BUG)
One in each engine accessory drive
Not speed regulated (not constant speed)
Output is regulated to 400 Hz by a converter
Come on line and stay on line slightly below idle engine RPM

Converter
Receives input from both backup generators at all times engines are running
Can select and process inputs from only one BUG or the other, not both simultaneously

Permanent Magnet Generators (PMG)
Two contained in each BUG
Supply AC power to flight control Power Supply Assemblies (PSA)
PSAs convert power from AC to DC for flight control computer use
Output is dedicated to flight controls—cannot be diverted elsewhere

DC Electrical System

Key DC Components Powered by DC

Main DC Electrical System
♦ PFDs ♦ NDs ♦ GPS ♦ ADIRU

Flight Control DC Electrical System

Transformer Rectifier Units (TRUs)

TRU	Powers/*Alternate For*	Backup From
L TRU	L Main DC Bus *L Flt Ctrl Bus* *R Main DC Bus*	Bus Tie to R
R TRU	R Main DC Bus *R Flt Ctrl Bus* *L Main DC Bus*	Bus Tie to L
C1 TRU	Captain's Flight Instrument Bus Battery Bus Hot Battery Bus Standby Inverter *Center Flight Control PSA* *FO Flight Instrument Bus*	Bus Tie to F/O *(Through Battery Bus)* *(Through Hot Battery Bus)*
C2 TRU	First Officer's Flt Instrument Bus *CA Flight Instrument Bus*	Bus Tie to CA

DC Busses

Hot Battery Bus—With battery switch off, powers 3 items
♦ Engine fire extinguisher bottles
♦ APU fire extinguisher bottle
♦ ADIRU, if switch is on

With only battery power available in flight, these systems are powered: Three items above, plus
♦ Standby Power System
♦ Captain's displays
♦ Left AIMS

Flight Control DC System—Designed to provide:
Dedicated and uninterrupted power to the primary flight control system PSAs

Primary power sources
Left and Right IDG Generators
Permanent Magnet Generators (PMGs)
Housed within backup generators
Dedicated power for flight control busses

Backup Power Sources
L and R Main DC Busses--can power alternate side Main DC Bus
Center Flight Control Bus
Alternate power—captain's flight instrument bus
Various power sources can power this bus

Hot Battery Bus can also provide power to left and center flight control bus

Individual Batteries
Connected to each flight control bus
Provide power for "short periods only" to insure no interruptions only during power transfers

Standby Electrical System

System Overview

Supplies AC and DC Power to selected key items

System Components
Main Battery
Standby Inverter
RAT Generator & associated generator control unit
C1 & C2 TRUs

Main Battery

Provides standby power to
♦ Hot Battery Bus ♦ Left and Center Flight Control Busses
♦ Battery Bus ♦ Captain's Flight Instrument Bus
♦ Standby Inverter, which powers the AC Standby Bus

Note: *Main battery can power the standby system for a minimum of 5 minutes..QRH Electrical 6.6*

Main Battery Charger
Powered by R Main DC Bus (through the Ground Service Bus)
When powered, the battery charger powers
Hot Battery Bus
L and Center Flight control busses

Battery Switch
ON—On the ground and not powered:
Some switch annunciator lights are illuminated
Allows the APU to be started.
On the ground after AC power is removed or lost:
Standby busses and emergency lighting are powered
The left inboard, outboard, and upper center displays, and the left CDU are powered.

OFF—On the ground, battery bus not powered and the EICAS advisory message **ELEC BATTERY OFF** is displayed
In flight, the EICAS advisory message **ELEC BATTERY OFF** is displayed.

Standby Inverter

Converts DC to AC
Powers AC Standby Bus if L Transfer Bus is not powered

Standby Power Switch

Note: *Ground operation only*
OFF—The AC Standby Bus is not powered
AUTO (Guarded)—Standby Busses transfer to battery power if normal AC power is lost
BAT (Momentary)—Standby Busses are powered from the battery if AC power is not available
Initiates a DC / standby self-test if AC power is available.

Ram Air Turbine (RAT)

Provides Hydraulic Power to Center system flight controls
See Hydraulic system, *page 51* for details function hydraulic

No limits on time or speeds for deployment

Provides standby electrical power for
C1 & C2 TRUs
Priority is hydraulic pressure over electrical power generation
Power for standby system switches to Battery if RAT sheds electrical power generation

Power from RAT both Captain's and F/O's flight instrument busses are powered through C1 & C2 TRUs

RAT Deployment switch
Press—deploys RAT
UNLKD Light (*Amber*) indicates RAT is not in stowed position
PRESS Light (*Green*) indicates RAT is deployed and producing ≥ 1500psi

RAT Deploys if: (HEED)

- **H***ydraulic—All 3 hydraulic system pressures are low* ***OR***
- **E***lectrical—Both AC transfer busses are unpowered* ***OR***
- **E***ngines both fail* ***AND*** *center hydraulic pressure low* ***OR***
- **D***eployed manually*

Acronyms

ELMS .... Electrical load management system
IDG .... Integrated drive generator
PSAPower Supply Assemblies
TRU .... Transformer Rectifier Unit

Battery Only—Key cockpit items shaped in a "T"

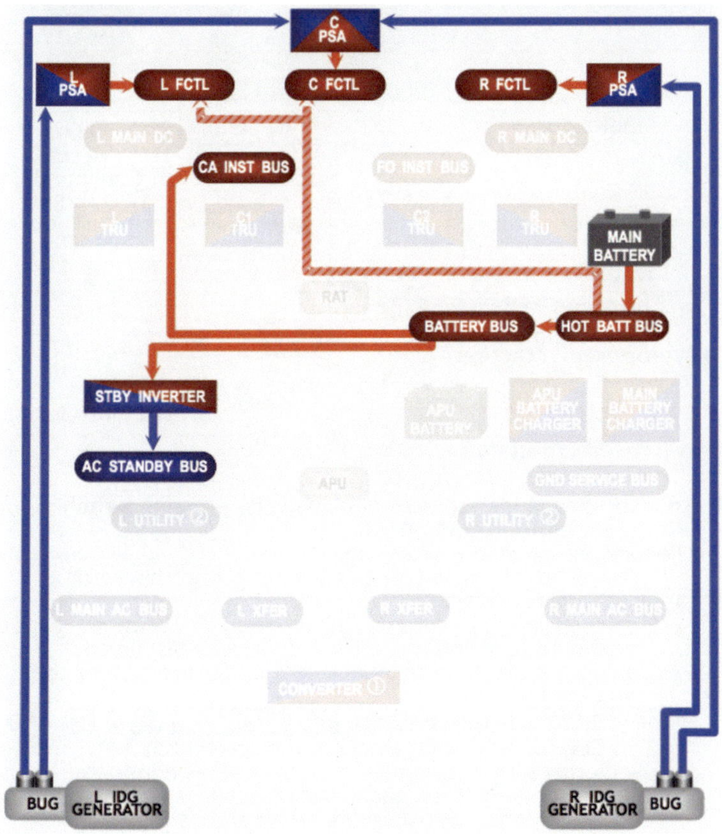

RAT Power plus Battery—Key cockpit items

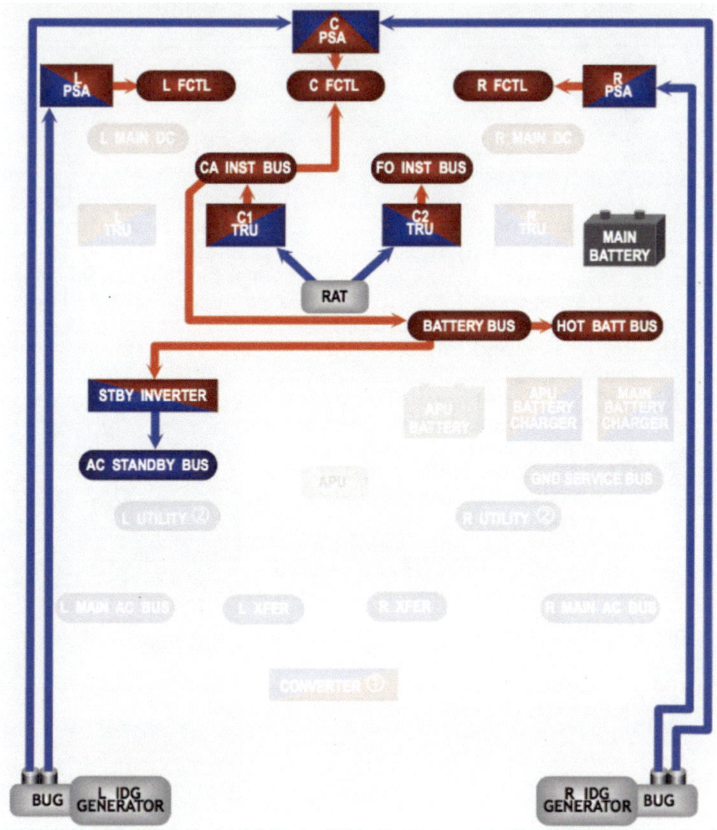

Engines

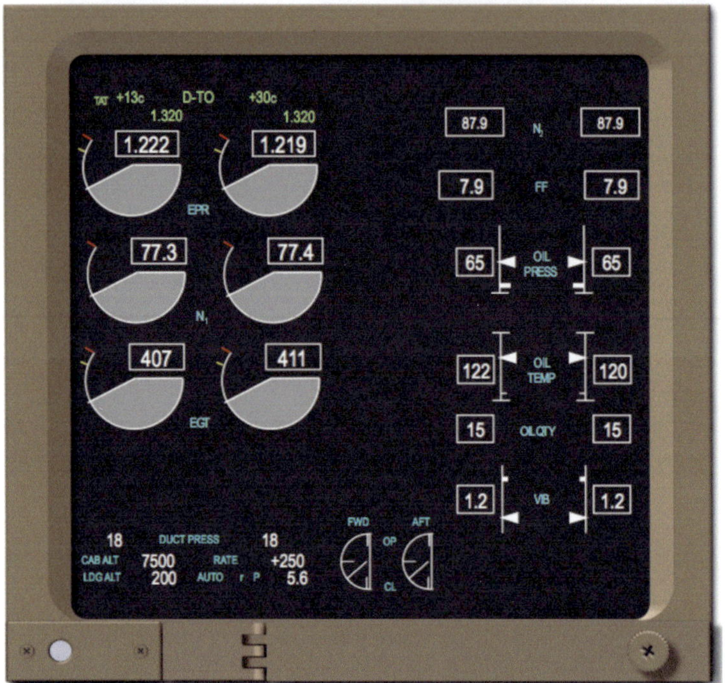

General Engine Characteristics

200 ER Designation ....Rolls Royce Trent 892
Thrust Rating (each)...........90,000 pounds at Takeoff Thrust
300ER Designation ....General Electric GE 90-115BL
Thrust Rating (each) ....115,000 pounds at Takeoff Thrust

Compressor-Turbines (3 Sections)

- N_1—Rotor: Fan and low pressure turbine section on a common shaft
- N_2—Rotor: Intermediate pressure compressor and turbine sections on a common shaft
- **200 ER** N_3—Rotor: High pressure compressor and turbine sections on a common shaft
- Three sections are mechanically independent (**300ER** two sections)
- N_3 Rotor drives engine accessory gearbox

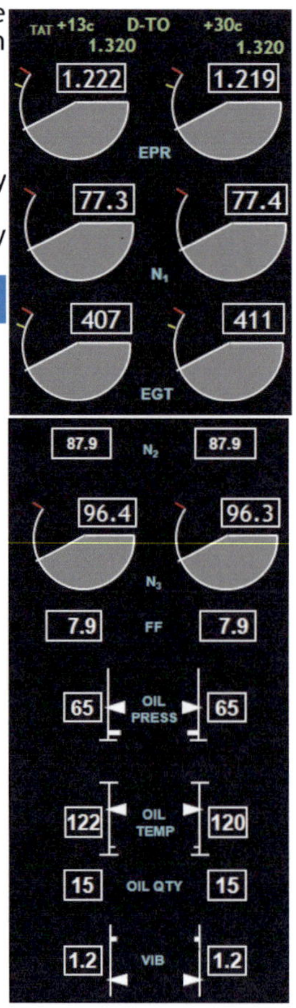

Engine Indications

Primary Engine Indications

- **200 ER** EPR, N_1, and EGT
- **300ER** N1 and EGT
- Displayed on EICAS by default

Secondary Indications

- N_2, N3, fuel flow, oil pressure, temperature & quantity, and vibration (**300ER** no N3)
- Displayed on MFD with DSP **ENG** button
- If MFD failure occurs, can display in compact format on EICAS
- Automatic display of secondary indications commanded when any of the following occurs:
  - ♦ Displays initially receive electrical power
  - ♦ Fuel control switch moved to cutoff in flight
  - ♦ Engine fire switch is pulled in flight
  - ♦ Secondary parameter limit is exceeded
  - ♦ N_3 RPM below idle in flight
- When these conditions occur, secondary display cannot be cleared until condition is no longer present

Abnormal Secondary Instrument Indications

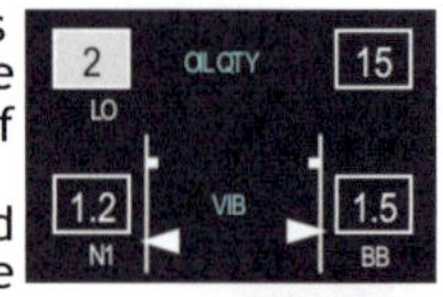

- Vibration Monitoring—indicates source (**N1**, **N2** or **N2**) or **BB** (broadband) if source can't be determined
- Oil quantity Low—**LO** displayed and numbers reverse to black on white when low

Exceedance Indication

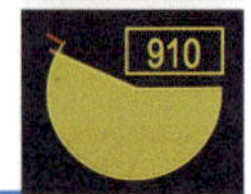

- If amber band or red line exceedance conditions occur, indication box and pointer turn to red
- When condition returns to normal range
  - Box remains red
  - Numbers return to white
- DSP **CANC / RCL** button returns box to white

Electronic Engine Controls (EECs)

Full authority over engine operation

- Uses thrust lever inputs to automatically control forward and reverse thrust
- Two control modes
  - Normal—Uses EPR as primary thrust reference setting
    - **300ER** Uses N1 as primary thrust reference control
  - Alternate—Uses N1 RPM to set thrust

Maximum rated thrust is available in any phase of flight by moving thrust levers to the full forward positions

Alternate mode—selected automatically

- Called "Soft" alternate
- Switch is still in normal
- A/T remains engaged
- N1 becomes primary thrust controlling parameter

	Normal	Soft Alternate	Hard Alternate

Indications	**ALL**	**EPR:** Blank **N1:** Actual Red Line	**EPR:** Blank **N1:** Actual Command Ref/Target Maximum Red Line
Thrust Ref N1	**300ER**	X	X
Thrust Ref EPR	**200ER**		
Thrust Limit (Over*boost*) Protection	X	**200ER** & **300ER** Available with A/T **300ER** Available at full forward thrust lever	
Over*speed* Protection	X	X	X

(X Indicates feature is available)

Hard Alternate Mode—selected manually by pressing switch

- No thrust limit protection
- Reduce thrust before selecting
- With autothrottle engaged, thrust limit protection returns

300ER EEC Differences

- All three modes—actual, command, reference/target, maximum, and red line N1 information is displayed.
- Automatic reversion or manual selection to alternate mode
  - Indicated by the
    - EICAS advisory message **ENG EEC MODE** (**L** or **R**) *AND*
    - EEC **ALTN** light on associated mode switch
  - Selecting alternate mode on both engines eliminates thrust lever stagger at equal thrust settings, or asymmetric thrust when thrust levers are operated together
  - Autothrottles remain engaged when EEC automatically switches to alternate mode.
  - FMC computes alternate mode N1 reference/target values
- EICAS Advisory message **ENG RPM LIMITED** (**L** / **R**) provided when overspeed protection is active

Note: *Autothrottles remains engaged in soft or hard alternate mode*

EEC Idle Selection
EEC Selects approach idle or minimum idle automatically
Higher thrust for quicker engine response is provided with approach idle selection
Approach idle is selected in flight if:
- ♦ Engine anti-ice is operating
- ♦ Flaps in landing position (25 or 30°)
- ♦ One hydraulic air-driven demand pump is INOP, and the flaps are out of the UP position
- ♦ The opposite engine bleed air valve is closed

AutoStart

Normal start system
Start selectors control starter air valves
Fuel cutoff switches control ignition and fuel flow
EEC monitors start and commands start selectors to **NORM** and bleed air cut off at 50% N_3
Packs are shut off when start cycle begins and are automatically restarted after start cycle is complete

Start is aborted for	Restart OK	Ground, No Restart
Hot Start	★	
Hung Start	★	
No EGT Rise	★	
Compressor Stall	★	
Starter Shaft Failure		X
***A**ir—Insufficient air pressure*		X
***N1**—No N_1 rotation*		X
***D**uty—Start time exceeds starter duty cycle*		X
★ *Restart attempts occur automatically for these items*		

AutoStart is not aborted for
Oil pressure or oil temperature
These are ***not monitored*** by autostart
In-flight assisted autostarts
System temporarily discontinues the start ***IF***
- ♦ T/O EGT red line limit is reached, ***OR***
- ♦ No light-off ***OR***
- ♦ Hung start is detected

For any of these, autostart system
- ♦ Windmills engine 30 seconds before next attempt
- ♦ Uses windmill motoring between attempts (starter is re-engaged on the following start attempt)
- ♦ Corrective action occurs automatically for unspecified "detected problems" without aborting start attempt
- ♦ During second and subsequent start attempts—autostart re-introduces fuel & ignition when EGT falls below 200°C

Thrust Control Malfunction Protection

Protects against high thrust malfunction on the ground
Shuts down engine when
- ♦ Airplane on ground ***AND***
- ♦ Thrust lever at idle ***AND***
- ♦ Engine above idle speed and not decelerating normally

EICAS warning message ENG FAIL (L or R) displays on PFD if
Thrust is less than commanded with
Airspeed between 65 knots and 6 knots prior to V1
EICAS caution message ENG FAIL (L or R) displays when engine speed reduces below idle speed
Remains displayed until fuel cutoff switch is in **CUTOFF**
EICAS caution Message ENG THRUST (L or R) displays when:
- ♦ Actual thrust is significantly less than commanded ***AND***
- ♦ Actual thrust is not increasing to commanded thrust, ***AND*** Airspeed is greater than 6 knots below V_1

Autostart

ON - Arms the autostart system.
OFF - Switch **OFF**, Autostart system is disabled
Start must be manually controlled

Manual Starts

Autostart Switch OFF selects manual start mode
Fuel control switch should not be moved to **RUN** with EGT over 100°C ....See OM Volume I, Supp Proc 45.1
OM Volume II, ENGINES, APU 7.20.12, 7.20.13

Engine Start Switches

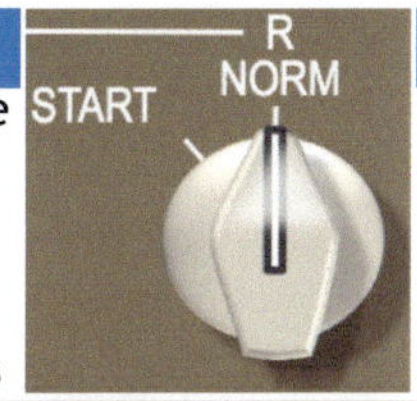

START—Initiates engine start by opening the start valve
Releases to **NORM** at start valve cutout
NORM—The start valve closes
300ER **CON**—Both ignitors operate continuously (if the **FUEL CONTROL** switch is in **RUN**) when:
- ♦ On ground, N1 > approximately 55% N1 with takeoff flaps set, ***OR***
- ♦ In flight

Engine Ignition

Two igniters
EEC Selects igniter for each start
Igniters are alternated on subsequent starts
In flight starts always use both igniters
Power sources
Normal—Main AC power
Backup—Standby AC (Standby inverter)

Auto Relight Function

Activates below idle with fuel control switches in **RUN**
Continues until engine spools down to down to 35% N_3
If system is deactivated due to N_3 below 35%
It may be reenergized by moving fuel cutoff to **CUTOFF** and back to **RUN**
This is the reason for the **Red Box** procedure
Flameout protection provided in heavy hail/rain
300ER Differences
EEC-Detected flameout—both ignitors activate
Sub-idle stall detection—fuel shut off for one second to clear stall then fuel restarts

Engine Fuel System

200ER **Fuel path to engine**
- ♦ Fuel flows through a <u>spar valve</u> located in the respective main tank
- ♦ <u>Engine stage fuel pump</u> increases pressure
- ♦ Preheating occurs in an <u>engine-oil heat exchanger</u>
- ♦ A <u>fuel filter</u> removes contaminants
- ♦ A <u>second fuel pump</u> further increases pressure
- ♦ <u>Fuel is metered</u> in accordance with commanded thrust
- ♦ Fuel flows last through the <u>engine fuel valve</u> before reaching the injectors

<u>Both fuel valves</u> close when
The fire switch is pulled ***or***
The engine fuel control switch is moved to **CUTOFF**
300ER **Fuel path slightly different**
Spar valve through <u>engine stage fuel pump</u> same as above, then:
- ♦ A <u>second fuel pump</u> further increases pressure
- ♦ Preheating occurs in an <u>engine-oil heat exchanger</u>
- ♦ A <u>fuel filter</u> removes contaminants

<u>Fuel metering</u> and remaining path to engine is the same as above for -200 series engines
Diagrams.............................AOM, ENGINES, APU 7.20.19

Fuel Control Switches

RUN (Autostart **ON**)—Opens spar fuel valve
Arms engine fuel valve (EEC opens valve when required)
Arms selected ignitor(s)
EEC turns ignitors on when required
RUN (Autostart **OFF**)—
Opens spar fuel valve
Opens engine fuel valve
Turns ignitors on
CUTOFF
Closes the fuel valves
Removes ignitor power
Unlocks the Engine Fire Switch

Engine Oil System

The engine oil system lubricates and cools
- Engine main bearings
- Gears & accessory drives

The system provides
- Fuel heating
- Fuel system icing protection

No Minimum oil quantity (amber or red line limit)
- Low oil quantity causes automatic display of secondary engine indications
- Reverses display to show quantity in black on white background

300ER Oil Filter Clogging
- If filter clogs, it is bypassed
- EICAS advisory message **ENG OIL FILTER** (**L** or **R**) is displayed

Thrust Reversers

Hydraulically operated
- Use system pressure from respective side hydraulic system

Levers can only be raised when
- On the ground
- Thrust Levers are at idle

When reverse levers are pulled aft to the interlock position:
- Autothrottle disengages
- Auto speed brakes deploy

When reverse is terminated, thrust levers cannot be advanced until reverse levers are fully down

EICAS advisories
- Advisory message **ENG REV LIMITED** (**L** or **R**)
  - Reverser cannot deploy when commanded or will be limited to idle
  - Not all conditions detectable
  - If other limiting conditions, thrust levers simply restricted to idle reverse
- Advisory message **ENG REVERSER** (**L** or **R**)
  - Displayed on ground to indicate reverser system fault

<u>Note</u>: 200ER *Following any rejected takeoff where thrust reversers were deployed, maintenance must determine if an inspection of the thrust reversers is necessary prior to the next takeoff.............QRH LANDING GEAR 14.6, BRAKE TEMP*

Engine Failure Alert

Triggered when engine produces less than commanded thrust

Message ENG FAIL is displayed (*Red*) on PFD if
- Actual thrust is less than commanded and
- Speed is between 65 knots and V_1

Message ENG FAIL L or **R** is displayed (*Caution*) on EICAS if engine decelerates to less than idle speed
- Remains displayed until:
  - Engine thrust recovers ***or***
  - Fuel control switch is moved to **CUTOFF**

Message ENG THRUST L or **R** is displayed (*Caution*) on EICAS if
- Actual thrust is significantly less than commanded ***and***
- Actual thrust is not increasing ***and***
- Speed is between 65 knots and 6 knots below V_1

APU—Auxiliary Power Unit General

Priorities
- Electrical power ***then***
- Bleed Air

Operating Envelope
- Can operate and provide electric power throughout aircraft envelope
- Bleed air only available to 22,000' pressure altitude

Starting

Two starting systems
- Air turbine starter
  - Uses engine bleed air or ground cart air
- Electric motor starter
  - Powered by APU battery
  - Main airplane battery powers
    - Inlet door
    - Fuel valve
    - Fire detection system
- Starter selection is automatic
- Bleed air start is used if available

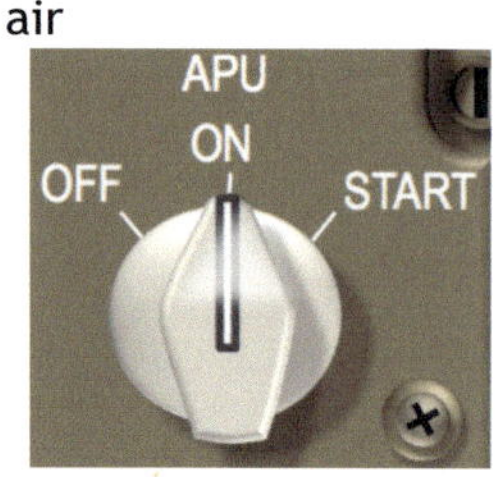

Fuel
- Supply is from left main fuel manifold
- Left forward AC fuel boost pump operates continuously to provide fuel if
  - AC power is available ***and***
  - APU Selector on
  - (*Regardless of boost pump switch position*)
- For APU starting and running when only DC power is available, a DC pump in the left main tank operates

APU Shuts down automatically if start attempt fails

In Flight Automatic Start occurs when Both AC transfer busses lose power
- APU Starts regardless of switch position
- May be shut down by moving selector to **ON**, then **OFF**

In-Flight Start Reliability Program
- Selected flights—crews requested to attempt in flight start after 2 hours at cruise altitude
- Maximum attempts if unsuccessful start—two additional attempts (three total attempts)
- Reliability improvement
  - Place Start Switch in **ON** for one minute before start
    - Allows start logic to complete before selecting **START**
    - Allows APU inlet door to fully open
  - Momentarily hold start switch in **START**

APU Shutdown

APU Cool down cycle begins automatically when selector is moved to **OFF**

Cycle lasts 105 seconds under a no-load condition

If APU Selector is moved back to RUN within the cool down cycle, APU stays running

APU must slow to below 12% before restart can be initiated

APU Operating Modes

Attended Mode—Definition
- Either engine running or starting ***or***
- Airplane in flight

APU Shuts down for any of these faults with no cool down period
- ♦ APU Fire or inlet over temperature
- ♦ Overspeed or loss of speed protection
- ♦ APU controller failure
- ♦ Speed droop

EICAS Advisory message **APU SHUTDOWN** displays

<u>Note</u>: *Previous OM versions describe a speed drop as RPM <88% and no acceleration for 10 seconds*

APU Continues to operate with these faults:
- ♦ High EGT
- ♦ High oil temperature
- ♦ Low oil pressure

Following these faults, there is no cool down when the APU is shut down

For any of these, EICAS caution message **APU LIMIT** displays

Unattended Mode Definition
- On ground, with no engines running or starting

Shutdown occurs automatically for any of the faults above, **plus:**
- ♦ Generator oil filter approaching bypass
- ♦ Intake door failure
- ♦ No combustion on start
- ♦ No acceleration on start

No cool down period for any of these prior to shutdown

EICAS Advisory message **APU SHUTDOWN** displays

APU Fuel Usage

Approximately 600 pounds per hour

See AOM 21c.2.2 APU Fuel Usage

Acronyms

EICAS ....Engine indicating and crew alerting system
EEC ....Engine electronic controller
EPR ....Engine pressure ratio
MFD ....Multifunction Display

Fire Protection

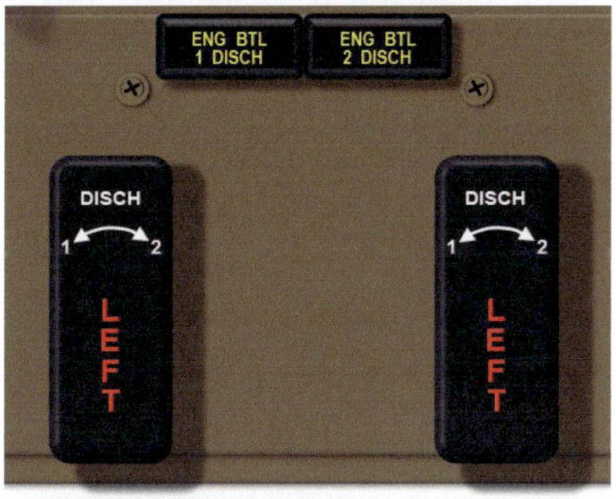

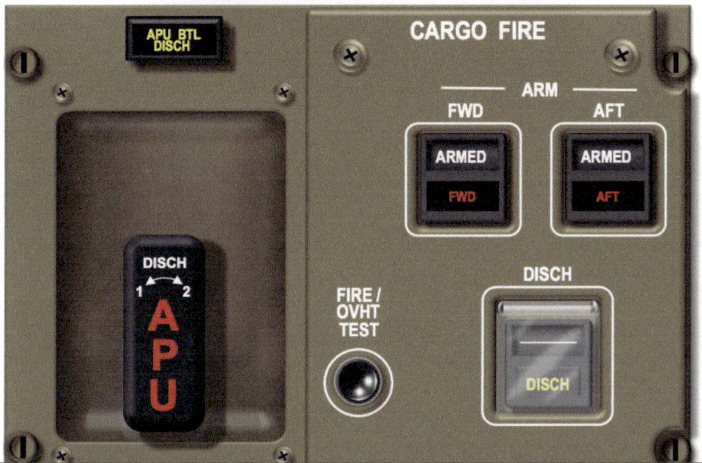

General

Fire Protection and Overheat Detection

Component	Fire Detection	Extinguishing System	Overheat detection
Both Engines	X	X	X
APU	X	X	
Cargo Compartment	X	X	
Lavatories	X	X	
FD Crew Rest Area	X		
Main Deck Crew Rest Area	X		
300ER Upper Deck Crew Rest Areas	X		
Main Wheel Wells	X		

Power Sources

Unlike other aircraft—Power sources for fire protection / detection not specified nor emphasized in ASM or in ground school

QRH does not include lists of inoperative systems

Sufficient redundancy exists to protect fire-related systems

Engine & APU Fire/Overheat Protection

Fire & Overheat Detector Loops

Dual loops on each engine, the APU, and in wheel wells
- Engines only—loops provide both fire and overheat detection

Normal operation—both loops must detect fire for warning activation

Loop Fault
- <u>Single loop failure</u>—Switch to single loop sensing occurs
  - Fire detected in remaining loop results in fire WARNING or overheat CAUTION activation
- <u>Double loop failure</u>—EICAS advisory message DET FIRE ENG L or R results

APU Fire Detection results in APU automatic shutdown
- If APU is unattended, single APU fire bottle also discharges automatically
- APU considered unattended with the airplane on the ground if engines are not running or starting

Fire Warnings

Fire detection results in:
- <u>Aural Warnings</u>
  - ♦ Fire warning bell (intermittent)
  - <u>Fire Warning Messages and Lights</u>
  - ♦ Both Master **WARNING** lights come ON
  - ♦ EICAS Warning messages
    - ENG FIRE L and ENG FIRE R
    - FIRE APU
  - ♦ Affected Fuel Control Fire light ON (*Red*)
  - ♦ Affected fire switch ON (*Red*) for engines or APU
  - <u>Mechanical Changes</u>
  - ♦ Affected fire switch unlock for engines or APU
- <u>APU Only</u>
  - ♦ APU Shutdown for APU fire (attended <u>***or***</u> unattended mode)
  - ♦ Nose strut mounted fire warning horn if on ground
  - ♦ Fire bottle automatically released if APU fire signal is generated with engines not running or being started (unattended mode)

Overheat Cautions

Aural Warning—Caution beeper sounds

Fire Warning Messages and Lights
- ♦ EICAS—<u>Caution message</u> OVERHEAT ENG L *or* R

Both Master CAUTION lights
- ♦ EICAS message remains on until overheat condition is terminated

Fire Bottles

Eight Total, / 300ER, Nine Total
- 2: Engines
- 5: Cargo compartments / 300ER, 6 in cargo compartments
- 1: APU

Two provided for use in engines
- Either or both can be discharged into either engine
- <u>Pulling Engine Fire Switch</u>
  - Arms discharge squib on each bottle
- <u>Rotating Switch</u>
  - Selects the bottle to be used ***and***
  - Discharges it into the associated engine
- <u>Rotating Switch in Opposite Direction</u>
  - Selects the opposite bottle ***and***
  - Discharges it into the same engine

One for APU
- Manual discharge
  - Pulling APU Fire Switch arms bottle for discharge
  - Rotating switch in either direction discharges bottle into APU compartment
- Automatic Discharge
  - Occurs with airplane on the ground and engines not running or being started
  - Discharge occurs a few seconds after APU automatic shutdown

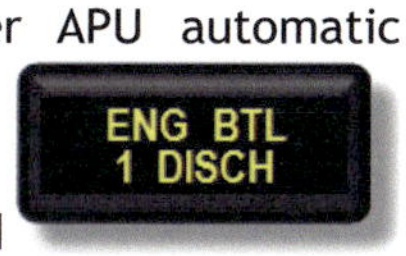

If a bottle discharges or has low pressure:
- **ENG BTL 1** or **2 DISCH** light illuminates
- EICAS Advisory message **BOTTLE 1** or **2 DISCH ENG**

Wheel Well Fire Protection

Dual detector loops

Main Wheel wells only (not nose gear well)

No fire protection provided

Lavatory Detection

200ER SMOKE LAVATORY EICAS advisory message displays

300ER SMOKE LAV/COMP EICAS advisory message displays

Cargo Compartment Fire Protection

Smoke Detection

Both forward and aft cargo compartments divided into three zones
If **smoke** detected in any zone, **fire warning** results for that compartment
If **detection system is inoperative**, EICAS advisory message results **DET FIRE CARGO FWD** or **AFT**
200ER Three detection zones
300ER Four detection zones

Cargo Fire Switches

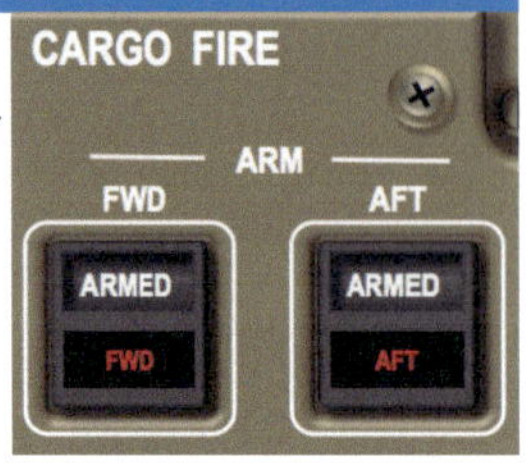

Armed (pushed)
- **Key Items** (*Arms*, *Aims*, *Air*):
  - *Arms* all cargo fire extinguisher bottles
    - *Aims*—Arms selected compartment extinguisher valve
  - *Air*—Reduced flow
    - ♦Turns off both lower recirculation fans
    - ♦Shuts down cargo heat
    - ♦Commands packs to low flow (minimum air flow required to provide pressurization)
    - ♦**300ER** Shuts down Nitrogen Generation System
- Aft cargo fire only:
  - Shuts down bulk cargo compartment ventilation system operation
  - Shuts down the lavatory & galley vent fan
- Forward cargo fire only: Puts the equipment cooling system into the override mode

DISCH Switch
- Discharges cargo fire suppression into armed compartment
- **DISCH** light in switch—fire extinguishers have discharged

Fire Warning

Indications of fire include:
- Fire bell
- Master **WARNING** light
- EICAS Warning message **FIRE CARGO FWD** or **AFT**
- **CARGO FIRE FWD** or **AFT** Fire warning light

Fire Extinguishing

200ER **Five fire bottles** provided
300ER **Six fire bottles** provided

In flight fire extinguishing
- Pressing the appropriate cargo fire discharge switch:
  - ♦ Activates a small explosive charge (squib)
  - ♦ Sound is a muffled explosive sound in the cabin
  - ♦ Two bottles are discharged fully into the affected compartment
  - ♦ After a time delay*
    - ♦♦If still airborne, remaining three bottles are discharged at a reduced rate
    - ♦♦If landing occurs, but aircraft is not shut down
      - Only 1 bottle is discharged at touchdown ***and***
      - Discharge is at a reduced rate

Ground fire extinguishing
- Pressing the cargo fire discharge switch:
  - ♦ Activates a small explosive charge (squib) as above
  - ♦ Two bottles are discharged fully into the affected compartment
  - ♦ After a time delay* if the aircraft is still not shut down, only 1 bottle is discharged (no metering)

***Note**: *Older versions of the operating manual identified the time delay as 20 minutes*

Fire System Tests and Controls

Fire / Overheat System Tests

Automatic
- Occurs at initial power up
- Continuous tests of loops occur while electrical power is applied
- Fault notification through EICAS is provided

Manual Test
- Accomplished by pressing and holding the Fire / Overheat Test Switch

Manual Test Indications:
Interior—1 Red EICAS message, 9 Red lights, fire bell
- ♦ EICAS **FIRE TEST IN PROG** ....1 Red
- ♦ Changes to **FIRE TEST PASS** *or* **FIRE TEST FAIL** when complete
- ♦ Master **WARNING** (2)...2 Red
- ♦ L and R Engine Fire Handles Light.......................2 Red
- ♦ APU Fire Switch Light ....1 Red
- ♦ FWD and AFT Cargo Fire Warning Lights...............2 Red
- ♦ Left and Right Fuel Control Switch Fire Lights.......2 Red
- ♦ Fire Bell .... Total: 10 Red

Exterior—1 Light, Fire Horn
- ♦ Nose Wheel Well APU Fire Warning Light..............1 Red
- ♦ Nose wheel well APU fire warning horn (on ground)

Engine Fire Handles

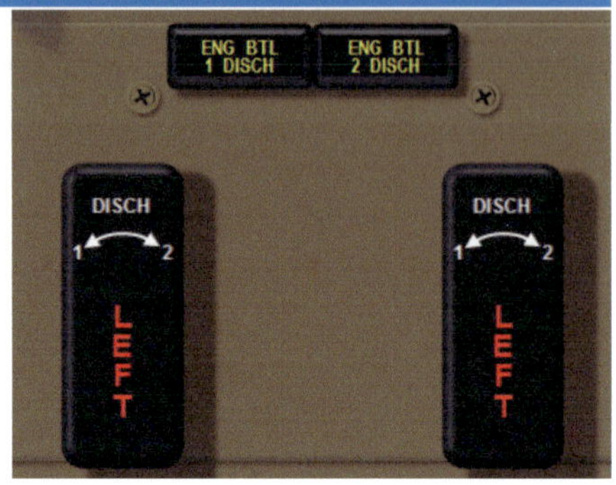

Normal position—in and locked
Switch unlock occurs for:
- Fire warning
- Fuel control switch in cutoff
- Override button is pressed

When pulled, the following actions take place:
- ♦ Arms engine fire extinguisher bottles
- ♦ Closes associated engine and spar fuel valve
- ♦ Closes associated bleed air valve
- ♦ Trips associated engine generator off
- ♦ Shuts off hydraulic fluid to engine-driven hydraulic pump
- ♦ Depressurizes respective engine-driven hydraulic pump
- ♦ Removes power to thrust reverser isolation valve

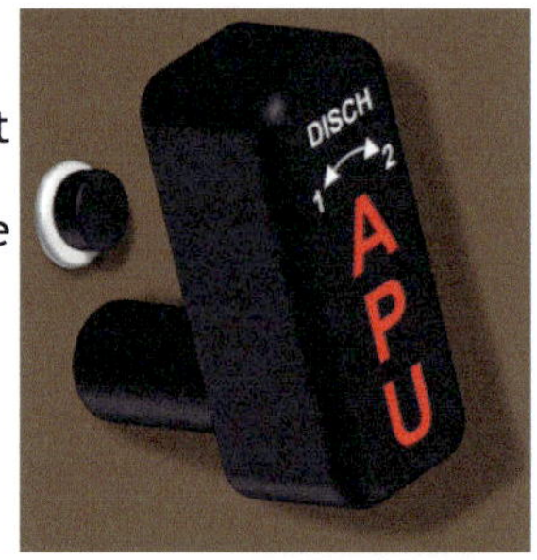

Rotated Left or **Right**—Discharges fire bottle 1 or 2 into respective engine
Engine Fire Switch Override Switch
- Button under switch
- Allows release of switch if electrical lock does not release

Engine bottle Discharged Light
- Indicates extinguisher bottle is discharged
- ***or***
- Low bottle pressure

Fuel Control Fire Warning Lights
- Associated fire detected ***or***
- Fire / Overheat Test Switch is Pressed

APU Fire Switch

Above applies to APU as well, except:
- Only one fire bottle
- Rotating switch either direction discharges single APU fire extinguisher bottle
- APU is shut down automatically on fire warning, not when switch is pulled
- Pulling the switch shuts down the APU if auto shutdown fails to do so for any reason

On the ground with both engines shut down
- APU fire signal from either fire detection loop causes
  - ♦ APU Shutdown ***and***
  - ♦ Extinguisher bottle discharge

Flight Controls

Flight Control Synoptic

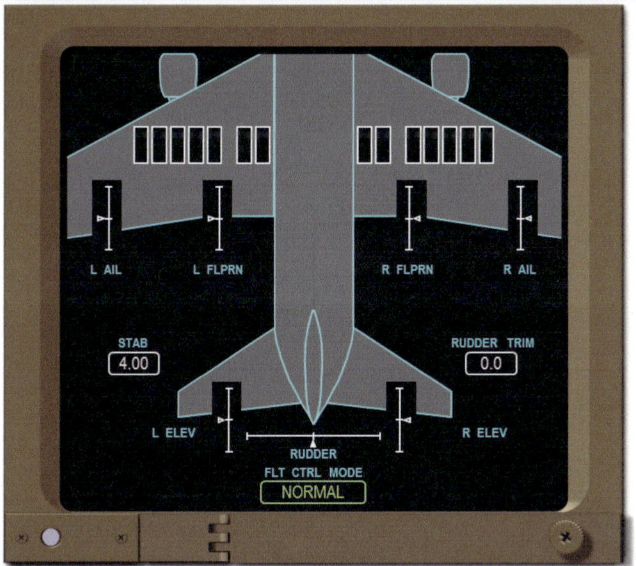

Flight Control Modes

Normal Mode

Designed mode for all operations

Flight control inputs command these system electronic components:
- Four actuator control electronics units (ACEs)
- Three primary flight computers (PFCs)

Manual Flight
- Pilot inputs sent to the four ACEs
- ACEs send these signals to the three PFCs
- PFCs verify signals and information from other airplane systems to compute control surface commands
- Commands are sent back to ACEs
- ACEs send enhanced signals to the flight control surface actuators

Autopilot Flight
- Autopilot sends signals to the PFCs
- PFCs generate control surface commands and send these to the ACEs
- ACEs then operate as above
- Signals are generated to move yoke and rudder pedals to reflect commanded motion of flight controls
- If pilot inputs override autopilot commands, A/P is disconnected and pilot inputs are used
- Autopilot flight only available in ***normal mode***

Secondary Mode

Entering Secondary Mode
- Occurs when PFCs can no longer support normal mode
- Causes may include:
  - ♦ Internal faults
  - ♦ Lack of required information from other airplane systems
  - ♦ Reversion to secondary mode then occurs automatically
- *Secondary mode cannot be manually selected*

Command to Flight Controls
- ACEs continue to receive pilot inputs on yoke and rudder pedals
- Signals sent to PFCs use simplified computations to generate flight control surface commands
- Simplified commands are returned to ACEs
- ACEs generate commands sent to flight control actuators

Flight Characteristics
- All flight controls surfaces remain operable
- Elevator and rudder are more sensitive "at some airspeeds"

Autopilot—Not available

EICAS Message
- **FLIGHT CONTROL MODE** Caution message and beeper
- Indicates flight controls are in secondary mode

Direct Mode

Entering Direct Mode
- <u>ACEs</u> transition to direct mode when
  - They detect the failure of all three PFCs ***or***
  - They lose communication with the PFCs
- <u>Manual selection</u> of direct mode
  - Primary Flight Control Computer Disconnect switch to **DISC** selects direct mode

EICAS Message
- **PRI FLIGHT COMPUTERS** Caution message and beeper
- Indicates flight controls are in direct mode

Flight Characteristics
- PFCs no longer generate control surface commands
- Pilot inputs are routed through ACEs directly to the control surface actuators
- Full airplane control is available for "continued safe flight and landing"
- Airplane handling qualities are "approximately the same as in secondary mode"

Mechanical Backup

Mode Design
- In event of complete electrical failure, direct connection to flight controls
- Cables from flight deck to stabilizer and two spoilers (4 & 11)
- Stabilizer control
  - Cable connected from alternate pitch trim levers mechanically connect to valves on each stabilizer trim control module
  - Hydraulic fluid directed to actuator to move stabilizer
- Spoilers
  - Control wheel mechanically controls spoilers 4 and 11
  - Direct cable connection from both yokes to actuators on these spoilers
- Yaw control
  - May be available through differential thrust

Flight Characteristics—Mode allows pilot to *"fly straight and level until the electrical system is restarted"*

Primary Flight Computer Disconnect Switch

DISC—Disconnects the primary flight computers (PFCs) from the flight control system
- Puts flight control system in the direct mode

AUTO—Normal position—Flight control system in **Normal** mode
- System faults automatically cause system to switch to secondary or direct modes.
- Reselection may allow restoration of secondary or normal mode operation

DISC Light—Indicates PFCs are disconnected (either manually or automatically)

Flight Mode Capabilities

Mode	Normal	Secondary	Direct
All normal & automatic features	X		
Yaw Damping	X	Some	
Manual Rudder Trim Cancel	X	X	
Rudder Ratio	X	Degraded	Degraded
Elevator Feel	X	Degraded	Degraded

Flight Envelope Protection

Stall Protection

Limits speed to which airplane may be trimmed
- At approximately minimum maneuvering speed, further trim inhibited toward slower speed (nose up)
- Pilot must apply back pressure to maintain slower speed
- Alternate pitch trim switches do not override this inhibit
- Flying near stall speed, aft yoke movement requires increasing force

Autothrottle Protection

<u>If</u>: .Autothrottle is armed and not engaged, ***and***
....Speed decreases to near stick shaker activation
<u>Then</u>: Autothrottle engages in the SPD mode ***and***
Advances thrust to maintain ***the greater of***:
Minimum maneuvering speed *(equates to near top of amber band)* ***or***
Speed set in MCP window
Autothrottle protection/engagement is inhibited:
Pitch control in TOGA or FLCH ***or***
Below 400' above airport on takeoff ***or***
Below 100' radio altitude on approach

Overspeed Protection

Limits speed to which airplane can be trimmed
Nose down trim is inhibited at V_{MO} / M_{MO}
Alternate pitch trim switches do not override this inhibit

Roll Envelope Bank Protection

Activates when airplane bank angle exceeds ≈35°, 40° and 45°
300ER Also sounds low altitude during takeoff and landing for bank angles as small as 10°
Result—Control wheel force rolls airplane back to within 30° bank angle
Limits on protection
Can be overridden by pilot
Full yoke deflection always results in full control surface displacement
Autopilot disengage bar disables bank angle protection
Indication—Excessive bank indicated on PFD bank indicator
Color changes to amber on both roll and slip/skid indications when 35° roll is exceeded.

See ASM 10.10.5 PFD Attitude Indications

Other Flight Control Features

Thrust Asymmetry Compensation

Engages when <u>thrust</u> differential between engines is ≥10%
Available at all times except when:
Airspeed below 70 knots on the ground ***or***
Reverse thrust applied ***or***
Flight controls are in secondary or direct mode ***or***
Automatically disengaged due to
System malfunction ***or***
Loss of engine thrust data
TAC Disengages when:
Engine thrust data is lost ***or***
Engine is damaged or surges
TAC may cause some rudder pedal deflection just before it disengages
Before liftoff—reduced compensation to allow pilot cues through roll and yaw

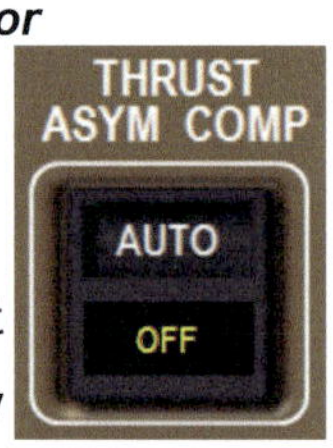

Wheel-Rudder Crosstie

Allows control-wheel only inputs to control initial effects of engine failure
Control wheel inputs deflect rudder up to 8°
Maintains sideslip capability consistent with crosswind landing requirements
Operation
♦ In flight only
♦ Speed < 210 KIAS
♦ Normal flight control mode only

Yaw Damping

Inputs rudder commands to accomplish three things
♦ Counteract tendency of swept-wing aircraft to Dutch roll
♦ Coordinate turns
♦ Gust suppression
Reduces effect of lateral gusts
Improves ride quality
No controls or indicators

Rudder & Aileron Trim Panel

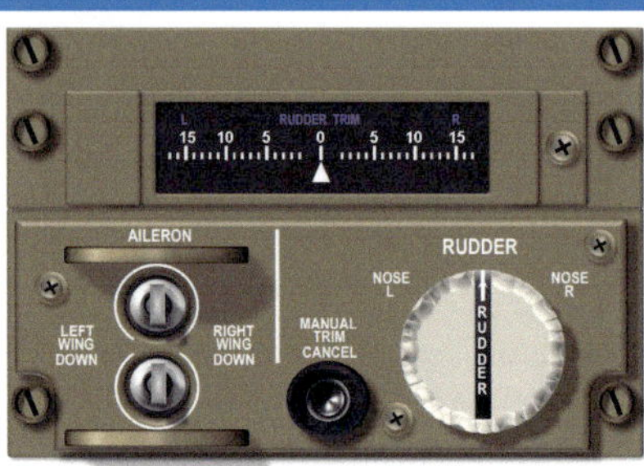

Stabilizer Trim

Actuation of both trim switches on either yoke:
Normal mode on ground—moves stabilizer directly
Normal mode in flight—changes trim reference airspeed
Secondary & direct modes—moves stabilizer directly
Stabilizer trim green band
Indicates allowable T/O trim range
Inputs—Gross weight, T/O Thrust, CG
If no inputs available from FMC—defaults to midrange
Stab signal invalid or not present—green band and pointer not displayed
Stabilizer Trim Cutout Switches
Respective C or R hydraulic power to the related stabilizer control module cut off

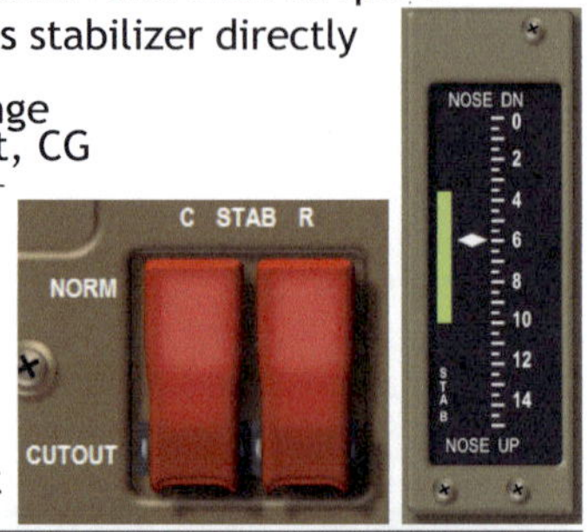

Flaps and Slats

Overview of Flap / Slat Positions

Flap Handle	Slats	Flaps	Range	Load Relief	
				Primary Mode	*Secondary Mode*
Up	Retract	Up			
1	Mid	Up			
5		5°	T/O		
15		15°		Flap Relief	
20		20°			
25	Full	25°	Land		Slat Relief
30		30°			

Flap and Slat Mode Comparison

Flaps and slats may be in separate modes
Flap Modes—PSA
Primary
Secondary
Alternate

Secondary Mode Entry

Occurs whenever primary mode fails to move slats or flaps to the selected position
Causes include:
Loss of center system hydraulic pressure
Flap component malfunctions
Restoration of primary mode
Restoration of center hydraulic pressure
Retraction of affected system surface(s) resets logic

Alternate Mode Entry

Occurs when Flap-Slat Electronic Units (FSEUs) are inoperative ***and***
Flaps and slats can no longer operate in the primary or secondary modes
Uses secondary drive motors
EICAS message **FLAP / SLAT CONTROL**
Manually entered with **ALTN** Switch
Arms alternate flap control mode
Arms Alternate Flaps Selector
Disables primary and secondary flap / slat mode operation
Inoperative:
♦ Flap Handle
♦ Flap / slat load relief
♦ Autoslat
♦ Asymmetry / skew protection
♦ Uncommanded motion protection

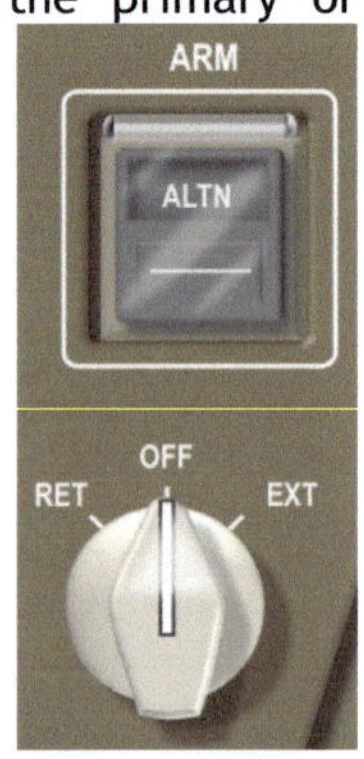

Flap / Slat Placard

On Landing Gear Panel

200ER

FLAP	LIMIT
1	255K
5	235K
15	215K
20	195K
25	185K
30	170K

300ER

FLAP	LIMIT
1	265K
5	245K
15	230K
20	220K
25	200K
30	180K

Mode capabilities

Mode	Primary	Secondary	Alternate
Activation	Normal operation	Automatically engaged	Manually engaged
Flap & Slat Positioning	Hydraulic	Electrical	Electrical
FSEUs	Control movement	Control movement	Bypassed
Indication	Flaps & Slats Primary 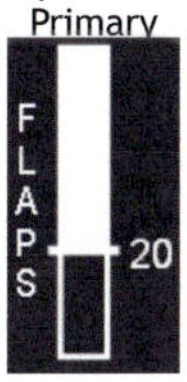	Slats Primary, Flaps Secondary 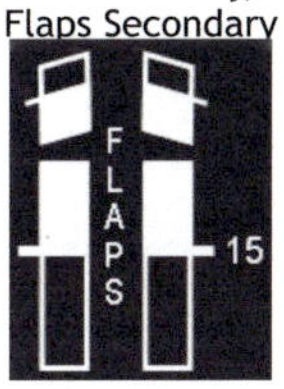	Flaps & Slats Alternate 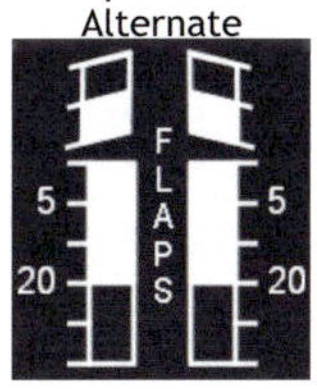
Protection Features	♦Autoslats	♦Slats to full down at flaps 1°	♦None
	♦Flap load relief	♦Slat load relief	♦Max flap extension 20
		♦If slats in mid range at point of failure, remain in mid range until flaps retracted to up or extended beyond 20°	♦Slats to sealed (midrange) ext.
	♦Asymmetry protection ♦Uncommanded motion protection		

Loss of position information—displayed as a white outline with no tape fill and no flap lever position indication

After normal retraction—10 seconds after full retraction, indication is removed from EICAS

Load Relief

Primary Mode—Flap Relief

Protects against excessive aero forces due to overspeed

With flaps in 15-30° settings, if placard speed exceeded

LOAD RELIEF Memo message appears

Flaps retract to appropriate setting for speed

Minimum retraction setting is 5°

When speed decreases, flaps return to lower position, (no lower than handle set position)

If flaps selected above placard speed

Flaps remain in safe range until speed decreases

Flaps then are lowered to selected range

Flap handle is not moved to reflect relief position

Secondary Mode—Slat Relief

If airspeed over 239 knots 200ER / 256 knots 300ER with slats down (gapped) ***OR***

Slats are retracted to midrange (sealed) position

LOAD RELIEF Memo message is displayed

Flaps must be retracted to full up to reset once this feature is activated

Inadvertent deployment—Cruise deployment above <u>altitude</u> approximately 20,000' is inhibited, ***OR*** <u>speed</u> 2bove

200ER 265 knots

300ER 275 knots

Landing Flap Settings

Normal—Flaps 25 or 30

Several abnormal procedures—flaps 20

Flaps 25 preferred normally due to:

- Fuel efficiency (saves ≈14 gallons of fuel per landing)
- Noise abatement
- Reduced flap wear
- During high (gusty) wind conditions.

<u>Note</u> *The 777 is not certified for flaps 25 autoland.*

Flaps 30 recommended when landing with:

- <u>**Runway**</u>
  - **Short** runway (less than 8000 feet)
  - **Slippery:** Braking action reported less than good
  - **Contaminated:** Wet / contaminated runway, or
- <u>**Weather**</u>: Tailwind
- **Deemed prudent** by the Captain.

AOM 9.1.3 Landing Flap Selection

Uncomanded Flap or Slat Motion

Uncommanded motion is detected when the slats or flaps:

- Move away from the commanded position ***OR***
- Continue to move after reaching a commanded position ***OR***
- Move in a direction opposite to that commanded

If the flap or slat is operating in the primary mode:

- Uncommanded motion causes automatic transfer to the secondary mode
- EICAS message caution FLAPS PRIMARY FAIL or SLATS PRIMARY FAIL is displayed.
- If motion continues, the system shuts down
- Resulting The EICAS message: FLAPS DRIVE or SLATS DRIVE

Spoilers

Location and naming

7 Sets of spoilers—(7 on each wing)

Numbered 1-14, L to R across aircraft wings

Symmetrically paired on opposing wings

Panels divided among the three hydraulic systems

EICAS advisory message SPOILERS indicates a pair has failed

Differential deployment aids roll control

Autospoilers

<u>Armed</u> when handle placed in **ARMED** position inflight

<u>Activated</u> when:

Gear on ground (not tilted) ***AND*** thrust levers both **IDLE**

OR

Either thrust lever moved to reverse **IDLE** detent

Speed Brake Use

Used to increase drag and decrease lift

Airborne use

Primary flight control mode—panels 5 & 10 locked out

In secondary and direct modes—panels 4, 5, 10 and 11 are locked out

Protects inboard ailerons

Speed brake deployment on landing

Automatic deployment occurs when the conditions for autospoiler deployment specified above are met

When armed, EICAS memo message, SPEEDBRAKE ARMED

Either thrust lever advanced to takeoff range causes handle to drive forward and retract speedbrake

Auto Slats

Stall warning system generates signal to autoslat system

Slats automatically extend

From mid-range, TO position (sealed)

To down (gapped) landing position

Slats retract "a few seconds after" signal is removed

System arming

Armed in 1, 5, 15 and 20° positions

Available only in

Primary slat mode ***and***

When slats are in the sealed position

Flight Control Hydraulic Power Shutoff Switch Panel

Overhead panel—Primarily for maintenance use

No pilot procedures use these circuit shutoff switches

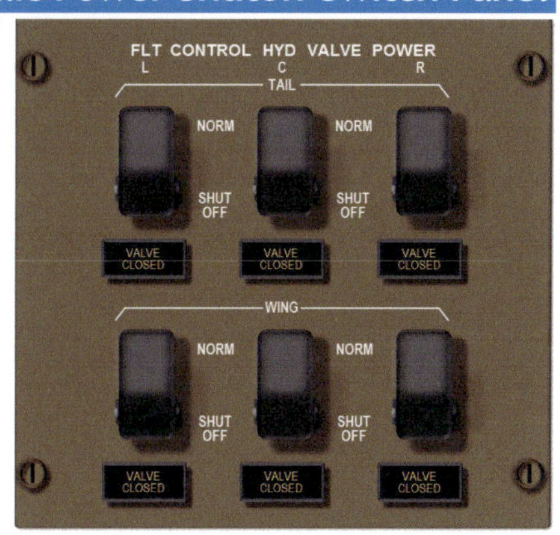

Flight Instruments

Note: *No attempt has been made here to completely describe the flight instrumentation displays and their connection to the FMS. These subjects are best learned using computer-aided color presentations, and are used on a day-to day basis by flight crews. Instead, key systems-related subjects have been summarized, which are easily forgotten and little used except for orals and recurrent training, and, of course, in the event of an actual malfunction. This section also serves as an overview for the transitioning crewmember.*

Displays

Six Liquid Crystal Display Units

Identical hardware
Display four primary types of information
- Primary flight display (PFD)
  - Displayed on outboard pilot station sides
  - Cannot be moved to other display units except in failure
- Navigation display (ND)
  - Normally displayed on inboard display positions
  - Can be displayed on lower center display
- Engine indicating and crew alerting system (EICAS)
  - Normally displayed on upper center position
  - Can be displayed on inboard or lower center positions
- Multifunction display (MFD)
  - Most commonly used position is lower center
  - Displays various information including
    - Electronic checklist
    - System synoptic
    - Communications information
    - Maintenance and other displays

Selection and Control
- ND information controlled by Electronic Flight Information System (EFIS) control panel on respective side
- If ND information is displayed on the lower center position, format is controlled by pilot whose inboard display position does ***not*** have ND displayed
- If neither pilot has navigation information displayed on their inboard display, then the left EFIS control panel controls the lower center ND information display
- If both pilots have ND displayed on inboard sides, and center ND is also displayed, left EFIS controls lower center panel and it is identical with left ND

Display Select Panel

Controls MFD format on left and right inboard and center lower display units
Upper buttons (green indicator lights) select where information will appear
Center two rows of buttons select system synoptic displays
Lower row of buttons selects checklist, communications, or navigation display
A second push of the same button blanks the active display
300ER **CAM** button is placed in blank position at right to activate ground maneuver camera system

New information replaces previous display if a new button is selected
If panel fails, displays can be controlled through any CDU
- CDU capability exists at all times
- If activated, disables respective EFIS panel
- Only one CDU at a time can control an MFD

If an inboard display selector is in EICAS position
- Only **ENG**, **AIR** and **FUEL** switches can affect the display
- These buttons cause compacted information for respective systems to be displayed on bottom of display

If a PFD (outboard) display fails
- PFD is displayed on respective side inboard position
- Display select panel cannot control that position
- Selection of information on respective DSP brings up information requested on lower center position

If Upper Center display fails
- EICAS information is automatically displayed on lower center position display
- If EICAS information is then selected to an inboard display position
  - Lower center EICAS displays secondary engine indications
  - Lower display can then be used normally as long as EICAS information is displayed on one of the inboard displays

Electronic Flight Information System (EFIS) Control Panels

Select display options, mode and range for respective PFDs and NDs
If panel fails, displays can be controlled through respective CDU
- Display with **<MENU> <DSP CTL>** entries
- CDU capability exists at all times
- If activated, disables respective EFIS panel

Display Brightness Control

Master brightness control provides simultaneous adjustment of all displays and control lighting
Individual brightness knobs
- Control brightness within a span when master brightness control is ***on***
- Control brightness in the full possible range for that display or panel with master brightness control ***off***

Instrument Display Source Selection

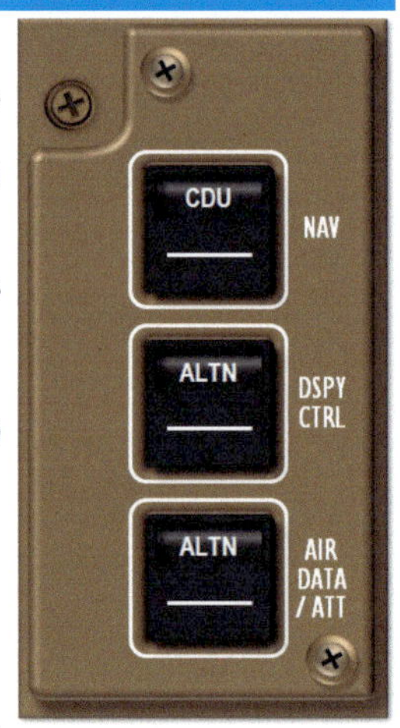

Display system automatically reconfigures for most failures
Source select switches provide alternate information sources for PFDs and NDs
- Automatic source selection occurs with switches off
  - This is switch out
  - **ALTN** and **CDU** switch annunciations not displayed

NAV Source Switch
- **OFF** (not depressed) Selects Auto
  - Active FMC provides data for ND map
  - ND Auto source—respective CDU
  - Right ND Auto
- **CDU** (**CDU** visible, depressed)
  - Selected CDU generates ND map

Display Control Source Switch
- **OFF** (not depressed) Selects Auto
  - Selects display processing channels
    - for respective side outboard and inboard displays
  - Automatically reconfigures channels as required for display processing channel failures
- **ALTN** (**ALTN** visible, depressed)
  - Selects alternate source for current display processing channel

Air Data/Attitude Control Source Switch
- **OFF** (not depressed) Selects ADIRU for air data and attitude information to the PFD and ND
  - Alternate sources or SAARU are automatically selected to replace ADIRU air data or attitude, as required
- **ALTN** (**ALTN** visible, depressed)
  - Selects alternate source for PFD and ND as follows:
    - Air data (CA side): ....SAARU single channel
    - Air data (FO side): ....ADIRU single channel
    - Attitude (both sides): ....SAARU

Undetected source failures (part of display missing or incorrect)
- Switches allow pilot to select alternate source manually

Center display control source switch is provided below center display
- Same logic as **DSPL CTRL** source switch is used for center upper and lower display units

Standby Flight Instruments

Standby Magnetic Compass

Liquid damped, standby compass on center upper windscreen brace

Heading correction information is on attached card

Clocks

Display airplane information management system (AIMS) generated:

- UTC time and date from GPS ***OR***
- Manually set time and date

Can also display

Alternating day and month-year

Elapsed time and chronograph functions

AIMS UTC time information comes from GPS

Integrated Standby Flight Display (ISFD)

The ISFD (All AA airplanes) displays:

- Attitude
- Airspeed
- Altitude
- ILS
- Magnetic heading information

Power source is left flight control DC bus

Data sources

Pitot-Static—Center

Air Data—Standby modules

Altitude, attitude, and airspeed are independent of ADIRU and SAARU values.

Attitude Reset Switch must be reset on the ground while airplane is stationary

Standby Attitude Indicator (*If installed*)*

SAARU attitude information is displayed

Bank indicator and **pitch scale** only

No FD display is possible

Standby Airspeed Indicator (*If installed*)*

Current airspeed displayed digitally in a box

Information calculated from two standby air data modules

One module for pitot

One module for static

Standby Altimeter (*If installed*)*

Displays altitude information from standby (static) air data module

Current altitude displayed digitally

Pointer indicates altitude in hundreds of feet

One revolution is 1,000 feet

***Many carriers**, including AA, have removed these displays and have only ISFDs.

Display System Information Sources

Air Data Inertial Reference System (ADIRS)

Information provided for:

Primary, secondary and standby air data

Inertial reference information

Major Components

Air Data Reference Unit (ADIRU) 1

Secondary Attitude Air Data Reference Unit (SAARU) 1

Air data modules 8

Static ports 7

Pitot probes 3

Angle of attack (AOA) vanes 2

Total air temperature (TAT) probe 1

<u>Note</u>: *Engine inlets each have independent TAT probes not discussed here*

Air Data Inertial Reference Unit (ADIRU)

Primary source for following information:

- Speed
- Attitude
- Altitude
- Inertial navigation position

Processes information measured by

Inertial gyros and accelerometers

Its own air data module inputs

AOA vanes

Other related systems

Described further in AOM 11.20.5 Air Data Inertial Reference Unit (ADIRU) and in the <u>FMS, Navigation section</u> of this Study Guide

Secondary Attitude Air Data Reference Unit (SAARU)

Secondary source for everything ADIRU provides

Processes information from its own air data and inertial reference sources

Transmits pitch and roll information to standby attitude indicator

Does not provide navigation position data

Described further in ASM, 10.20.17 Display System Sources, p. 10-20-14

Heading Data Selection

Heading Reference Switch selects either magnetic (NORM) or true heading

Selected heading displays on

- PFDs
- NDs
- FMCs
- AFDS

True is normally selected

- North of 82°N
- South of 82°S
- In vicinity of Magnetic poles—as depicted on charts

Inboard Display Selector

PFD—Displays PFD on inboard display and blanks outer display

NAV—Displays ND and inhibits DSP selections to inboard display

MFD—Allows display of DSP-selected display

EICAS—Displays EICAS

Inhibits most DSP selections except limited FUEL, ENG and AIR displays

Blanks upper center display

Air Data

Three static ports on left, **four** on right side of airplane

Left and right static information is paired through pneumatic tubing to each of L, C, and R air data modules

Air data modules convert static air pressure to digital output used by various systems

Center static ports also connected to

- Independent air data module <u>to provide static pressure to</u>
- Standby airspeed indicator <u>and</u>
- Standby altimeter

Pitot Probes

R & C probes on right side of aircraft forward section

L probe on left side of aircraft forward section

Each probe provides air source for respective air data module

Air data modules process dynamic pressure information to digital output for other systems' use

Center probe also provides dynamic pressure to standby air data module

Angle of Attack (AOA)

Two vanes, one on each side of forward fuselage

ADIRU provides primary data source for

PFD AOA indicators

SAARU is backup source

If primary source fails, secondary backup selection is automatic

AOA signal generation is independent from pitot static system

Damage to nosecone and/or disruption of airflow over AOA vanes renders information unreliable

Total Air Temperature (TAT)

Probe mounted on forward fuselage left side

Temperature sensed by probe is used by ADIRU and SAARU to compute total air temperature

Static Air Temperature (SAT)

SAT displayed on CDU—**PROG** page 2 comes from ADIRU

If ADIRU data is invalid, SAARU computed value is used

Anti-Ice of Air Data Components

Occurs automatically

All components in this section are heated at all times when either engine is running ....ASM, ANTI-ICE 3.20.9 Probe Heat

Primary Flight Displays

Integrated display of all flight parameters

Failure Flags displayed for system failures
Displayed information is removed
- If source fails ***OR***
- No system source information is available

Normally displayed information is replaced by dashes
- No valid information is available
- May be due to out-of-range or malfunctioning nav aids

Airspeed Display

Tape and digital window
Mach displayed when mach is greater than 0.40
Airspeed trend vector indicates predicted airspeed in 10 seconds
Selected airspeed is displayed above tape

Vertical Speed

Displayed to right of altitude tape
Digital display appears when VV is greater than 400 fpm
Selected vertical speed bug appears in AFDS V/S mode

Altitude Display

Altitude tape on right side of PFD
Digital altitude readout in box on center of tape
If meters selected on EFIS panel
- Current altitude in meters shown above altitude window
- Altitude in feet remains
- Selected altitude is displayed at top of tape in meters

Selected altitude from MCP window is displayed above altitude tape
- Boxed when approaching altitude
- Selected altitude is also displayed on "bug" on altitude tape
- Does not reflect VNAV parameters which may inhibit climb or descent to MCP window altitude
- <u>Note</u>: *This can be very confusing when an altitude selected and displayed on the PFD has not become active*
  - *Key is bug position—if bug is aligned with current altitude, then selected altitude from MCP is not where the airplane is going until the pilot changes something*

Landing altitude reference bar displayed on inner edge of altitude tape
- 1000-500 feet above touchdown, tape is white
- 500-0 feet above touchdown, tape is amber

QNH and QFE settings are available
- AA uses only QNH
- PFD Altitude tape background color green indicates someone has left the mode in QFE

Attitude Indication

Fairly intuitive—only selected information presented here
Small rectangle under bank pointer indicates slip and skid conditions
Pitch limit indication is displayed
- At low speeds with flaps up
- Any time flaps are down

Steering Indications

TCAS Resolution Advisory steering appears in attitude information area
Flight Path Vector (FPV) symbol
- Represents airplane flight path angle vertically ***and***
  - Airplane drive angle laterally
- Appears when EFIS FPV switch is on ***or***
  - MCP FPA reference switch is on

Flight Path Angle (FPA) symbol
- Shows selected flight path angle
- Appears when MCP FPA reference switch is on ***and***
  - Either FD or AP is engaged

Display of FPA and FPV is larger and brighter if FD is off

Radio Altitude

Displayed when radio altitude sensors detect airplane is less than 2,500 feet above source
Always on—will appear in cruise flight when another airplane passes less than 2,500 feet below

Angle of Attack Indication

Upper right corner of PFD
Digital and analog displays
Shows:
- Stick shaker point
- Zero reference line—4:30 position
- Additional reference lines every 5 degrees from -5 to +20

Data is computed—not raw data

Engine Fail, GPWS, and PWS Alerts

Displayed in large capital letters between attitude display and heading/track compass rose
- Refer to................ASM 7.20.24 Engine Failure Alert System
  - ASM 15.10.4 Alerts on PFD
  - ASM 15.20.5 Aurals, Master WARNING/CAUTION Lights

Navigation Displays (ND)

Only selected information for review is presented. Most information is well covered in simulators and other training devices. Only easily forgotten and occasionally asked-for information is contained in these notes.

Position Trend Vector

Three dashed lines
Represent predicted position at 30, 60, and 90 seconds
- Based on bank angle and ground speed
- For ranges greater than 20 miles ....3 segments
- For 20 NM range ....2 segments
- For 10 NM range ....1 segment

VNAV Path Pointer
Displays vertical deviation from VNAV computed path
Appears when airplane first gets within ±400' from path
Only appears in descent
Digital display appears when more than 400' from path

TCAS Indications

See Warning Systems notes, *page 58*

CDU L, C, R ND Source Indication

Indicated map source (Green)
Displays ND Source if
- CDU is selected on navigation source select switch ***OR***
- Both FMCs fail ***OR***
- A manually selected FMC fails

ILS Park Annunciation

Indicates ILS receivers are untuned
Displayed at power up until
- 200NM from touchdown ***OR***
- Halfway to destination, whichever occurs first

Remains in view until receivers are tuned

Acronyms

ADIRS................................air data inertial reference system
ADIRU.................................air data inertial reference unit
ADM ....air data modules
AIMS ....airplane information management system
CDU ....control display unit
DSP ....display select panel
EFIS.............................electronic flight information system
FD. ....flight director
FPA ....flight path angle
FPV..flight path vector
LNAV ....lateral navigation
MFD...multi-function display
ND ....navigation display
PFD ....primary flight display
PWS ....predictive wind shear
SAARU ....secondary attitude air data reference unit
VNAV...vertical navigation

Flight Management, Navigation

This section contains only selected notes specific to the 777. Many aspects of navigation are common to other aircraft with which pilots are familiar.

Navigation System Inputs

- ♦VOR ♦ Global Positioning System (GPS)
- ♦DME ♦ Air Data Inertial Reference System (ADIRS)
- ♦ILS ♦ Flight Management System (FMS)
- ♦ADF ♦ ATC Transponder

Global Positioning System (GPS)

FMC Computes position

Uses inertial reference inputs continually
Uses GPS for position updates to fine-tune position

Annunciation on ND is GPS when FMC uses GPS inputs

Two independent GPS receivers
GPS tunes automatically

If ADIRU becomes inoperative

EICAS displays message **NAV ADIRU INERTIAL**
FMC uses only GPS data to navigate

GPS Inoperative

EICAS displays message **GPS**
Indicates both receivers Inoperative ***OR*** Unavailable

GPS Updating

GPS position updates should be used
- For all approaches in which the FMC database and approach procedures are referenced to the WGS-84 reference datum
- For all other operations unless a specific state requires other update provisions within their airspace (eg, to accommodate non-WGS reference datum or other reason).
- At all Company scheduled and alternate airports:
  - GPS position updating is allowed for all approaches
  - Unless specifically prohibited, GPS updating should remain on for all phases of flight

GPS updates should be inhibited:
- For terminal operations, if the airspace, FMC database, and charts are not referenced to the WGS-84 datum (unless other appropriate procedures are used).......AOM p.21j-41

Inhibiting GPS data inputs

Can be accomplished with **GPS NAV PROMPT** on **POS REF** page of FMC (Page 3/3) ....See ASM 11.20.3 Global Positioning System (GPS)

Inertial System

Air Data Inertial Reference System (ADIRS)

Calculates flight parameters to include
- ♦Position ♦ Altitude
- ♦Speed ♦ Attitude

Provides this data to
- ♦Displays ♦ Flight Management System
- ♦Flight Controls ♦ Engine Controls

Major components are ADIRU and SAARU modules

Air Data Inertial Reference Unit (ADIRU)

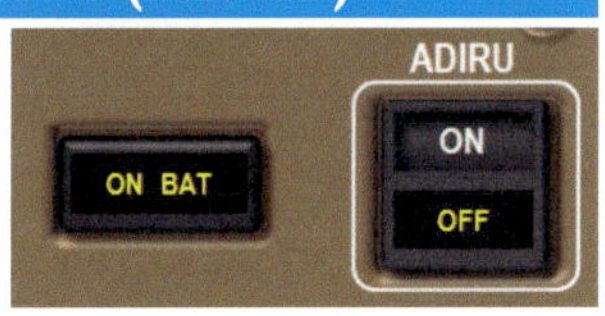

Provides primary
- Flight data
- Inertial reference
- Air data

Power
- Initial power-up requires battery power, and ADIRU switch on
- If ADIRU is turned off, full realignment required before aircraft is moved
- **ON BAT** Light—illuminates when
  - ♦ **ADIRU** is selected **ON *and***
  - ♦ Ground power or primary power subsequently is removed or fails
  - Hot battery bus continues to supply power
  - Ground crew call horn in nose gear well calls attention to ADIRU on battery power
- When ADIRU is switched off
  - Ground—Remains powered for "a few seconds"
  - Flight—System logic prevents ADIRU from losing power

Alignment
- Time to alignment complete is on ND upper left corner
- Airplane should not move during alignment time
- Position must be confirmed / entered
  - If entered position differs significantly from position of origin airport, scratchpad message ....**INERTIAL / ORIGIN DISAGREE**
  - If internal ADIRU comparison tests do not pass, scratchpad message.........................**ENTER INERTIAL POSITION**
  - If entered position fails comparison test twice, scratchpad message ....**ALIGNMENT REINITIATED**
  - If ADIRU position is "far different" from GPS position, EICAS message ....**NAV UNABLE RNP**
- Automatic realignment mode—velocity and acceleration errors are reset to zero
- No in-flight ADIRU realignment is possible
- Position realignment recommended if total time in navigation mode exceeds 24 hours

Secondary Attitude and Air Data Reference Unit (SAARU)

Secondary data source of critical flight data for displays and other systems

If ADIRU fails
- SAARU automatically supplies
  - ♦ Attitude ♦ Heading ♦ Air data
- In non-polar region, magnetic heading must be periodically updated

If ADIRU Inertial data fails
- EICAS caution message **NAV ADIRU INERTIAL**
- SAARU provides
  - Attitude data to PFDs
  - Three minutes of heading data in non-polar regions
  - Heading based on ADIRU heading prior to failure
- Control Display Unit (CDU) prompt after three minutes: **SET HDG**
- Heading must be "periodically updated" thereafter without further prompts
- SAARU heading continues to be provided until aircraft decelerates below 100 knots
- Numerous navigation modes and functions inoperative in this backup mode—see ASM for specifics
- Some capabilities are restored after SAARU heading update
- Autobrakes are inoperative
- No **LNAV** or **VNAV** mode capability

Power up
- Completely automatic at airplane power-up
- No SAARU controls available to crew

SAARU is sole source of data to standby attitude indicator display (*if installed*)

Air Data and Attitude Sources

Three Channel System

Left, center and right pitot and static systems

ADIRU and SAARU share data from these channels
- Air data modules—remote sensors for air data functions
- Both ADIRU and SAARU receive data from
  - *All* air data modules
  - Both AOA vanes
  - Dual element TAT probe

Air data is considered valid for ADIRU and SAARU when two or more channels agree

When ADIRU air data is invalid ***and***
- **AIR DATA SOURCE** switch is in **OFF** position,
- Valid SAARU air data is used

Single channel operation
- Occurs when ADIRU and SAARU data are invalid
  - Left PFD displays—**ADIRU** data from **L** pitot static system
  - Right PFD displays—**SAARU** data from **R** pitot static system
- EICAS message **NAV AIR DATA SYS**
- Left PFD receives data from L pitot static system

Standby Instruments
- Receive data from center pitot-static ports ***through***
  - Standby air data modules
- Attitude and airspeed displays are independent of ADIRU and SAARU values

Radio Navigation Systems

Automatic Direction Finding (ADF)

Two systems
Either ADF may be manually tuned from either left or right CDU
ND data displays ADF when related VOR/ADF switch in **ADF** position
ADF data displays in **CYAN**
If both FMCs fail, L and R ADFs may be tuned on related L and R CDU **ALTN NAV RADIO** page

Distance Measuring Equipment (DME)

Two installed
Normally auto-tuned
FMC tunes DME as needed for radio position updates
Can be manually tuned on FMC nav radio page
Note: *Manual DME tuning does not inhibit FMC DME tuning*
FMC uses two DMEs and triangulation to update position

Instrument Landing System (ILS)

Three receivers
Usually tuned automatically by FMC when
ILS, LOC back course, LDA or SDF approach selected for the active route
Runway is within
50 NM from the top of descent ***OR***
150 NM of the landing runway threshold, ***OR***
FMC is in **DESCENT** mode
Initial takeoff—ILS autotuning inhibited for 10 minutes to prevent PFD clutter
Selection and execution of an approach on active route during inhibited period overrides this limit
Inhibition does not apply on subsequent takeoffs on same flight (touch and go, etc.)
ILS Tuning Inhibit active when
AP is engaged and localizer or glideslope is captured ***OR***
FD is engaged, and either localizer or glideslope is captured with airplane below 500' radio altitude ***OR***
On ground, localizer is alive, airplane heading is within 45 degrees of localizer front course with airspeed greater than 40 knots
ILS Auto tuning is re-enabled when
Either **TO/GA** switch is pressed ***OR***
The autopilot is disengaged and both FDs are switched off ***OR***
MCP **APP** mode is deselected above 1500' radio altitude
Automatic retuning occurs when
ILS auto tuning is enabled ***AND***
A new approach is selected on the CDU
After Dual FMC failure
L and C ILS receivers are tuned with the L CDU
R ILS receiver is tuned with R CDU
Accomplished on the CDU **ALTN NAV RADIO** page
Can be manually tuned

NAVAID Identifier Decoding

Morse coded identifiers of tuned signals are automatically converted to alpha characters
Decoded identifiers appear on PFD
Verifying identifier with approach procedure is adequate for identifying correct navaid
Audio tuning is required if identifier does not display correctly as expected
Decoding applies to VOR, ILS, DME, and ADF

VOR Receivers

Two installed
FMC can tune VOR and collocated DME for position updates
FMC uses VOR / DME radio updates when more accurate sources are not available
Manual tuning
Accomplished by entering identifier or frequency
Inhibits automatic tuning until manual station is deleted
After Dual FMC failure
L and R CDUs tune respective VOR receiver
ALT NAV RADIO page of CDU is used
ND displays of VOR
Normally FMCs both tune same VOR
VOR bearing information appears in map display when
EFIS **VOR/ADF** switch is in **VOR** ***AND***
Either or both **VOR L** or **VOR R** switches are selected
VOR course deviation is displayed only in EFIS **VOR** mode

Transponder

Panel controls
Two ATC transponders
TCAS system

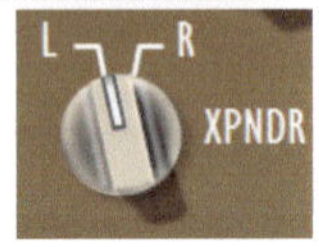

EICAS Advisory message
TRANSPONDER L or **R**
Indicates selected transponder failure
If altitude reporting fails, alternate source can be selected
Transponder information sources

NORM Altitude source selected—ADIRU
ALT Altitude source selected—SAARU
Therefore, **NORM** is always selected on preflight
See Warning Systems section, ASM for further information

Vertical Navigation

Note: *VNAV is adequately taught except for occasional hot items. These are gradually being added to this Study Guide.*

On Approach

VNAV On Approach mode is a non-annunciated mode which is important to understand
When active, allows the following:
IAS/MACH window can be opened and the command speed set while VNAV remains in VNAV PTH descent
MCP Altitude can be set above A/C altitude for missed approach
VNAV remains in VNAV PTH and follows descent path unless:
- Airplane accelerates to within 5 knots of current flap placard ***and***
- Airplane rises 150 feet above path

In this case, mode changes to VNAV SPD
Glide path angle, if specified on one or more legs of the approach procedure, is displayed when the mode becomes active
Becomes active when any of the following occur:
Airplane is in descent mode ***and*** flaps are out of UP
A VFR approach has been created and incorporated into the active flight plan ***and***
- The airplane has sequenced the FAXXX fix ***or***
- The airplane is enroute to a direct-to or intercept-to the RWXXX waypoint and the airplane is within 25NM of the runway threshold

A published instrument approach has been selected and incorporated into the active flight plan ***and***
- The airplane has sequenced (passed) the first waypoint on the published approach ***or***
- The airplane is enroute to a direct-to or intercept-to waypoint (1L on RTE page) and the airplane is within 12NM of the runway threshold ***or***
- The airplane is enroute to a direct-to or intercept-to waypoint (1L on RTE page) *that is the last waypoint in the approach* (runway or missed approach point) and the airplane is within 25NM of *that waypoint*

Flight Management System (FMS)

Components

Flight management computer (FMC)

- Two independent computers
- Normally, one active, with the other backing it up and prepared to activate if failures occur
- Inputs
  - Crew-entered flight plan data
  - Airplane systems data
  - FMC navigation database
- Outputs
  - Calculated present position
  - Pitch, roll, and thrust commands to fly optimal profile
  - Commands sent to APs, ATs, and FDs
  - Map and route data sent to NDs
- MCP used to select A/T, AP, and FD modes

Control Display Unit (CDU)

- Three mechanically identical units
- L and R are used for pilot navigation and some communication inputs
- C is used for communications and maintenance functions
  - SATCOM, flight interphone selections
  - Can be used to back up L and R CDU in event of failure
  - Can be used to select alternate display controls
- **300ER** Brightness adjustment is a rocker switch

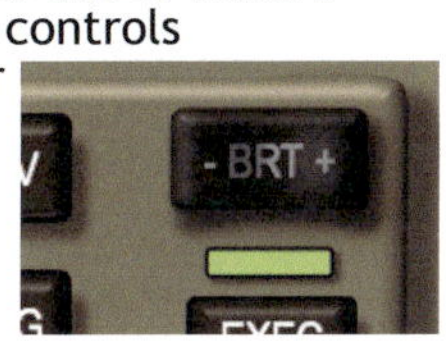

Acronyms

ADIRS .... air data inertial reference system
ADIRU .... air data inertial reference unit
CDU .... control data unit
DSP .... Display selector panel
FMC .... Flight management computer
FMS .... flight management system
GPS .... global positioning system
MCP .... mode control panel
MFD .... Multifunction display
ND .... navigation display
PFD .... primary flight display
SAARU .... secondary attitude and air data reference unit

Fuel System

General

Provides pressurized fuel to engines and APU
Fuel contained in center, left and right wing tanks

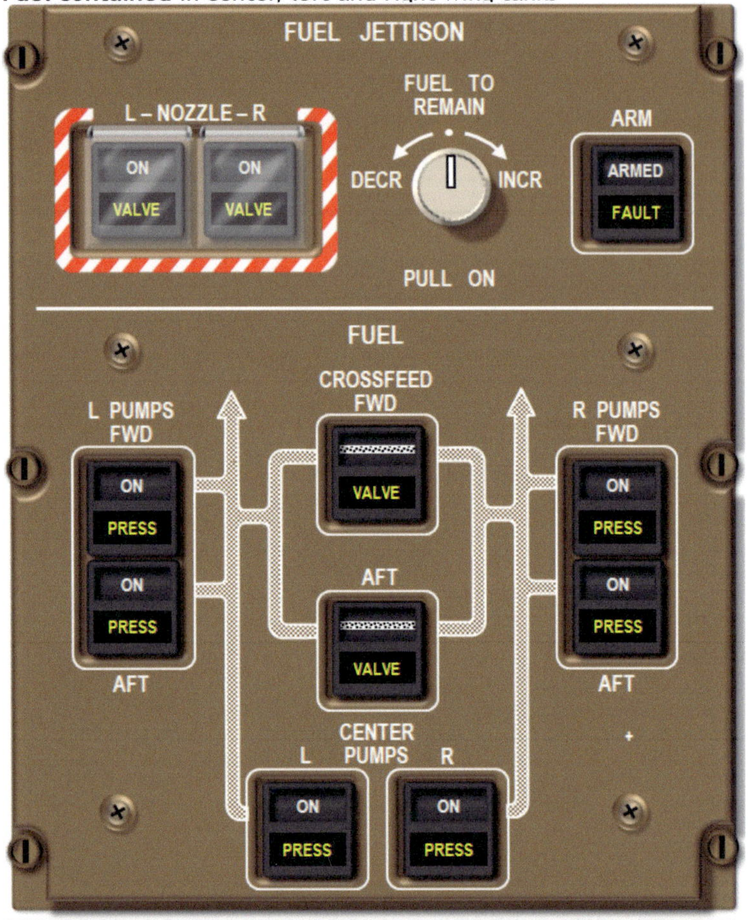

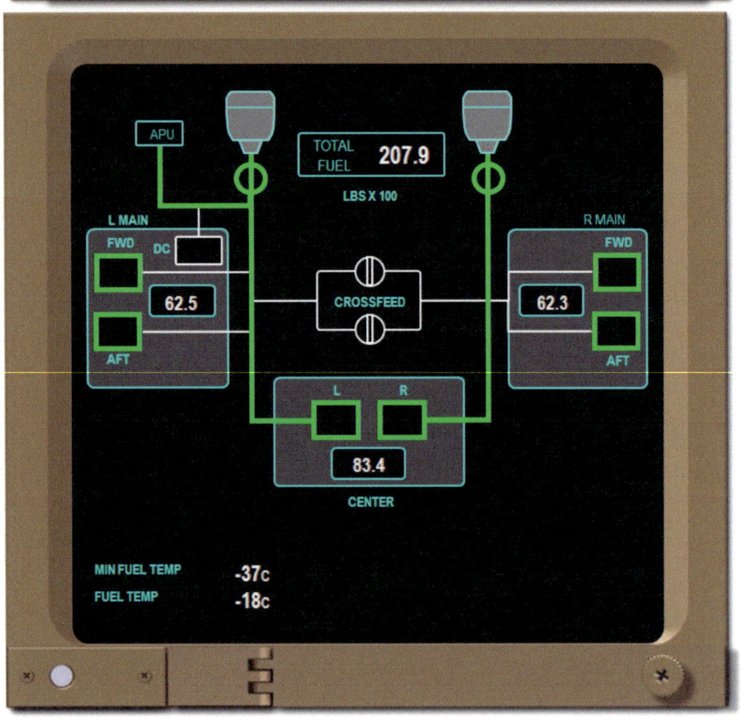

Fuel Measuring

Quantity
- Sensors in each tank
- Total fuel on primary EICAS display
- Individual quantities displayed on fuel synoptic

Temperature
- Displayed on primary EICAS display
- Normally white; Amber when approaching minimum allowed
- Minimum fuel temperature can be changed on FMS CDU
- **TO REMAIN** amount replaces temperature display when dumping
- Actual and min temperatures always displayed on fuel synoptic

Fuel Pumps

Two AC pumps in each tank
Any single pump has enough output to supply fuel to operate one engine under all conditions
Center tank pumps
- Override/jettison pumps
- Higher output pressure than wing tank pumps
- Allows center tank fuel to be used before wing tank fuel
- Low output pressure with more than 2400 lbs of fuel remaining FUEL PUMP CENTER L or R advisory message displays
- Low output pressure with less than 2400 lbs of fuel remaining FUEL LOW CENTER EICAS advisory message displays

Scavenge system
- Automatically transfers remaining fuel from center tank
- Process begins as wing tank quantity reduces (either L or R) to less than 29,000 lbs. 200ER or 52,600 300ER
- Center tank must be <35,000 lbs. (both 200ER and 300ER

Load shedding
- On ground, only one center tank pump operates unless two AC sources are available
- With only one AC power source
  - Only one pump operates
  - **PRESS** light on non-powered side is illuminated
  - Pump pressure EICAS message is inhibited

Indications with pump switches OFF
- Main tank pumps—**PRESS** light in switch and EICAS message
- Center tank pumps—Switch **PRESS** light and low pressure EICAS message inhibited

Left Main Tank DC Pump

Provides pressurized fuel to APU automatically when:
- APU Selector is on and
- No AC power is available

No controls for pump
Only indicator is on fuel synoptic
In flight operates automatically with loss of all AC to allow for:
- Quick left engine relight following loss in flight
- APU start

Also used on ground for battery start of left engine

Suction Feed

Can provide fuel to respective engine from main tanks
Designed to work if main tank pump output pressure is low
Suction feed lines bypass pumps
Dissolved air
- Can cause air bubbles as pressure lowers in climb
- Bubbles dissipate after stabilizing at cruise altitude
- Depletion time depends on altitude, temperature & fuel type

Suction feed is capable of providing sufficient fuel to operate engines after dissolved air depletes

300ER Nitrogen Generation System

Converts bleed air to nitrogen-enriched air
- Reduces flammability potential

Active during all phases of flight and for a brief period after landing
System is shut down when:
- ♦ Single engine operations *OR*
- ♦ Equipment cooling switch **OFF** *OR*
- ♦ Cargo Fire Arm switch is **ARMED**

Fuel Filters

Low Pressure Fuel Filter—between first and second stage of engine fuel pump
- Pressure drop between filter inlet and outlet sets **ENG FUEL FILTER** (**L** or **R**) status message
- Single channel indication system—so message can mean indication failure
- If the filter subsequently is blocked, a bypass line allows fuel to continue flowing to secondary pump

High Pressure Fuel Filter—Strainer downstream of second stage pump, fuel metering and flow meter
- Prevents blockages at engine combustor fuel spray nozzles
- No bypass or indication system
- Note: *High pressure fuel filter information removed from OM*

Crossfeed

Allows pressurized fuel from one main tank to feed both engines
Prolonged operation causes progressively larger fuel imbalance
Can also be used to balance fuel in main tanks
Crossfeed manifolds arranged so any pump can supply both engines (may not be sufficient for max fuel flow in all conditions)
- Crossfeed valves closed in normal operations
- Tanks then operate in isolation with one center pump and main tank pumps providing fuel to respective engine

Valve position disagreement with valve results in
- Crossfeed switch **VALVE** light ***AND***
- EICAS advisory message **FUEL CROSSFEED FWD** or **AFT**

Fuel Imbalance

If quantity disagrees between left and right main tanks by an "excessive quantity"
- EICAS advisory message **FUEL IMBALANCE**
- Balance attained through crossfeed
- Amount required to turn message on varies with total main tank fuel quantity
- See LIMITATIONS for allowable imbalance (or *page 11* of this study guide)

Imbalance pointer appears if main tank fuel differs L to R by more than 1000 pounds
- If a crossfeed valve is open, pointer appears if difference is over 200 pounds
- If message **FUEL IMBALANCE** is present, triangle becomes amber and is filled rather than empty white triangle
- Imbalance pointer flashes if balance is getting larger rather than smaller
- When fuel back in balance, **FUEL BALANCED** replaces **FUEL QTY** on expanded display for 5 seconds and flashes

Fuel Tank Capacities

200ER Nominal values	Gallons	Pounds
Left or Right Main	9,560	64,000
Center	26,100	174,900
Total	45,200	302,900

300ER Nominal values	Gallons	Pounds
Left or Right Main	10,300	69,000
Center	27,290	182,800
Total	47,890	320,800

For variations due to density, see..12.20.10 Fuel Tank Capacities or *page 11* of this study guide

Synoptic Fuel Spar Anomaly

L AC bus power interruption can cause invalid spar valve display
- When this occurs, fuel synoptic should not be used for situational awareness.
- Synoptic page on the CA inboard, FO inboard and lower MFD may show different indications

Indications:
- Fuel Synoptic page shows fuel spar valve as invalid (gray circle)
- Green fuel flow bars replaced with white background

APU Fuel Feed

Fuel supplied from left fuel manifold
Sources
- Any main or center tank pump which is pressurizing left fuel manifold
- DC pump in left main tank if no AC power is available to power another pump
- With AC power available and all boost pump switches **OFF**
  - Left forward pump operates automatically ***AND***
  - DC pump turns off

In flight with loss of all AC power, DC pump turns on automatically
- Allows for quick left engine relight
- Also allows for start of APU with no AC power

Jettison System

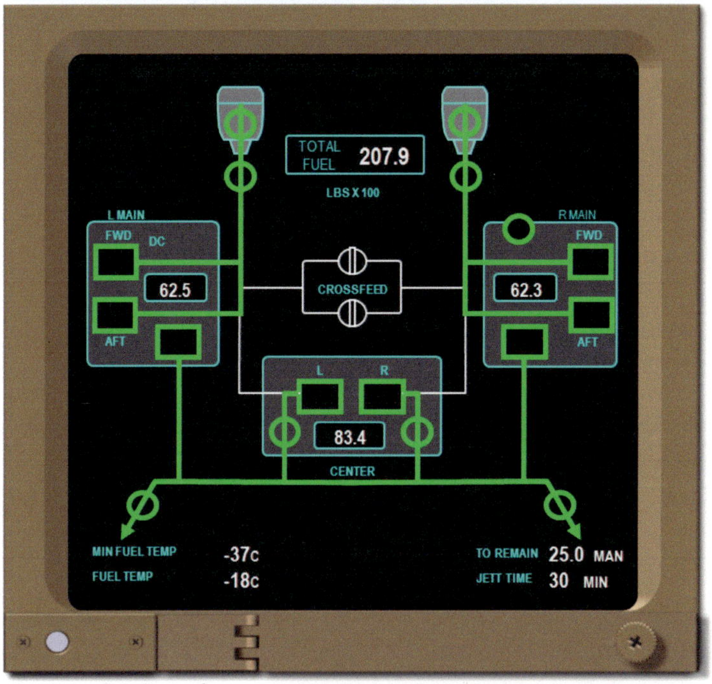

Jettison begins when:
- Fuel tank jettison **ARMED** switch is **ON** ***AND***
- Fuel nozzle switches are **ON** ***AND***
- Jettison nozzle valves are **OPEN** ***AND***
- Main tank jettison pumps operate

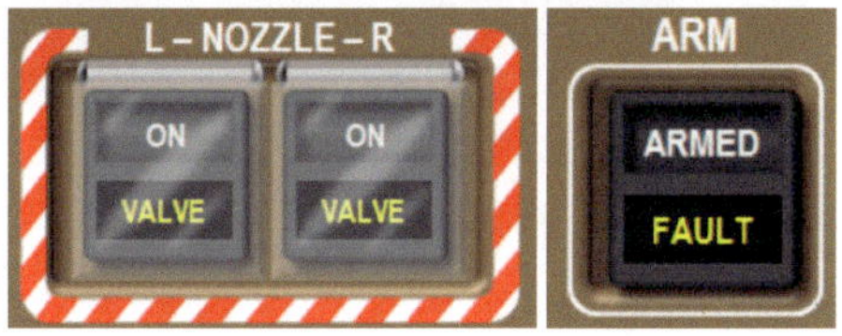

Jettison terminates when:
- Fuel quantity total reaches the value set in the fuel to remain display ***OR***
- Fuel jettison nozzles are both closed ***OR***
- Fuel jettison arm switch is disarmed ***OR***
- A fault is sensed in the jettison system ***OR***
- Fuel quantity in each main tank decreases to **200ER** 11,500 lbs. / **300ER** 8,500 lbs. ***OR***
- Aircraft is on the ground

Jettison Sequencing
- All tanks jettison simultaneously unless CG stability requires Center tank to be delayed in dumping
- Fuel synoptic is modified to show dump status, including valves and pumps

Jettison Progress
- Fuel Synoptic displays pump status and fuel flow to wing nozzles
- Expanded fuel EICAS display shows both time remaining and the amount of fuel scheduled to remain when dump is completed

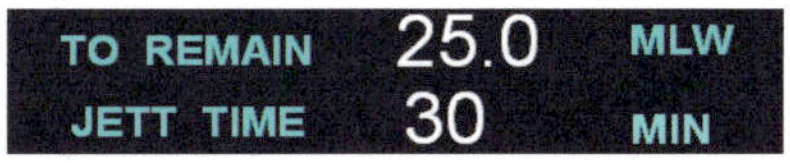

- When parameters above are reached, jettison terminates

Dump Rates
- Fuel in center tank (as well as wing tanks) ....5,400 PPM
- Fuel only remaining in wing tanks ....3,100 PPM

QRH FUEL 12.2, FUEL AUTO JETTISON

Hydraulics

General Features

Three independent systems—Left, right, center

Used to control

- Flight controls
- Leading edge slats
- Trailing edge flaps
- Landing gear
- Wheel brakes
- Nose and main gear steering
- Thrust reversers

Flight controls distributed among three systems
Any one system provides adequate aircraft control
Reservoirs for each system pressurized by bleed air

Left & Right Systems

Each system powers

- Flight control components
- Respective thrust reverser

Difference—Right system powers normal brakes as well
Primary pumps—Mechanically driven
One on each engine
Demand pumps
Electrically-powered
Supplementary pressure for high-demand situations
Backup for primary pump failure
Auto position—pumps run when:
Takeoff & landing (flaps extended)
System or primary pump pressure drops
On ground—right pump operates continuously

Center System

Key components powered

- Flight controls
- Alternate brakes
- Reserve brakes *
- Landing Gear actuation
- Leading Edge Slats
- Trailing Edge Flaps
- Main gear steering
- Nose gear steering

*Note: *Removed from Landing Gear description of brake system, but still appears in Hydraulic System section of ASM*
Ram Air Turbine (**RAT**)—Backup pressure for primary flight controls only

Primary Pumps

Two electrically powered pumps
Designated C1 and C2 pumps
Load shedding
Ground with only one AC source

- C1 pump runs at all times when selected
- C2 pump does not run if C1 is running

With two AC sources (including primary and secondary external AC) both pumps run
In flight C2 may load shed by ELMS in following conditions:

- All other electrical pumps are running *OR*
- Only one AC source is available *OR*
- Generator capacity is exceeded

When conditions for load shedding no longer exist, pumps are powered again automatically

Demand Pumps

Air driven
Supplementary pressure for high demand
Backup for primary pump failures
Controlled by C1 and C2 demand pump selectors
AUTO position—pumps run if:

- System and/or primary pump pressure low *OR*
- System logic anticipates high demand

ON position—pump(s) run continuously
Both pump selectors **ON**—only C1 operates
Both pumps cannot run simultaneously
Demand Pump Fault Lights display for

- Low demand pump output pressure
- Excessive demand pump fluid pressure
- Demand pump is selected **OFF**

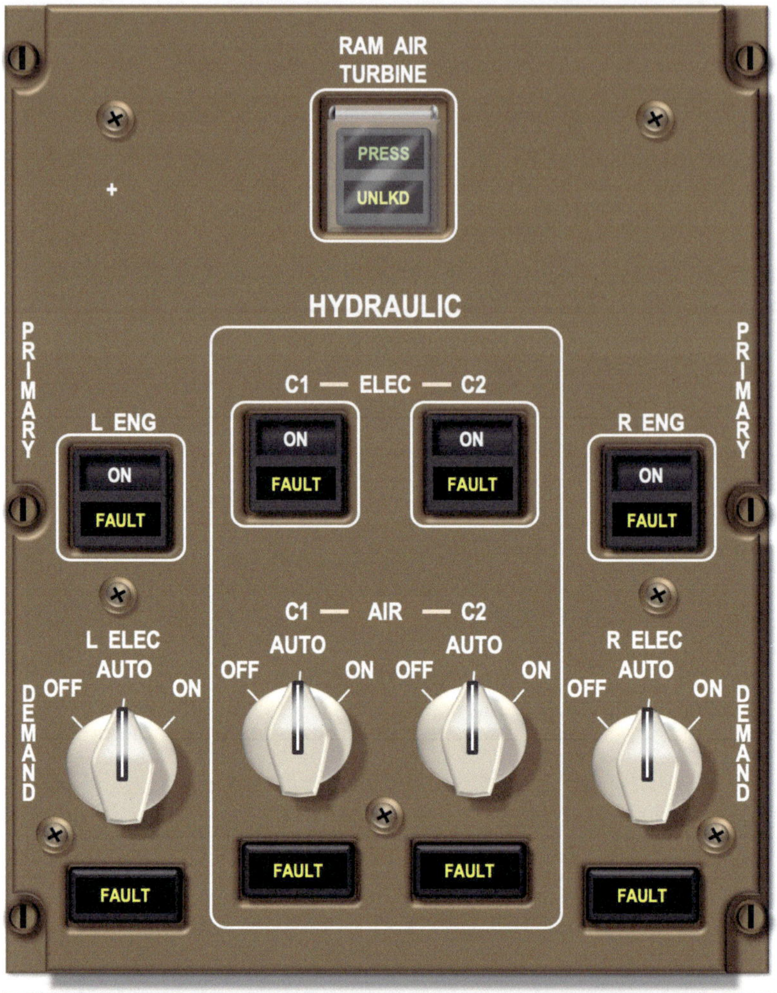

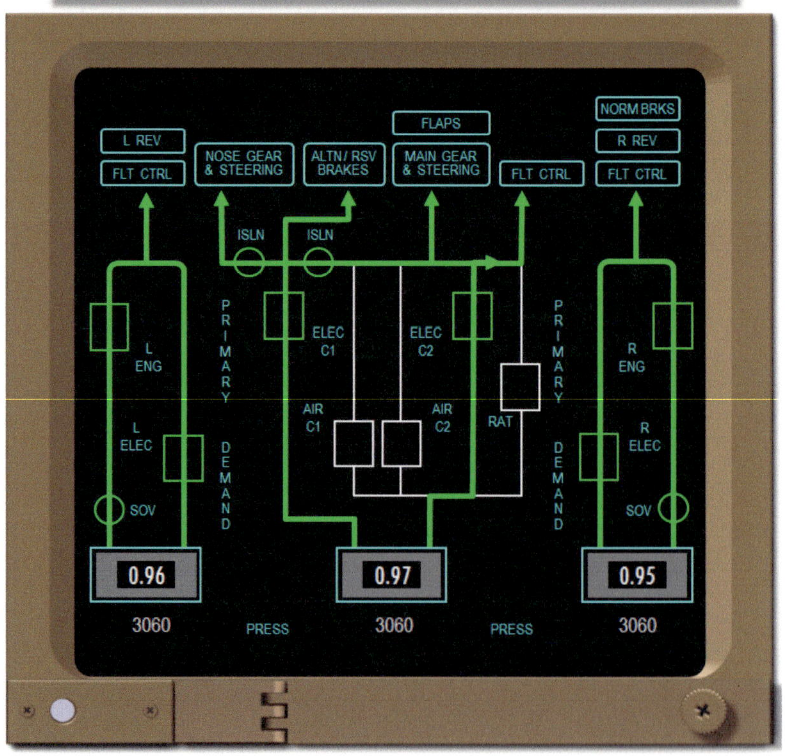

Synoptic Indications

Quantity—reservoir volume as a percentage of normal service level (1.00 is the normal service level) 0.95
- Reservoir quantity low .... 0.15 LO
- Quantity over full (inhibited in flight)........................ 1.05 OF
- Quantity requires refilling (Inhibited in flight) .... .72 RF

Pressure of highest output pump pressurizing system .... PRESS 3010

Synoptic graphic symbols

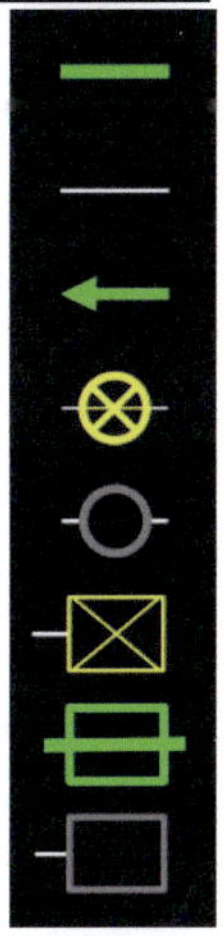

- Pressurized line..
- Unpressurized line ....
- Flow direction ....
- Hydraulic valve not in commanded position ....
- Valve with invalid sensor data ....
- Pump commanded on with pressure low ....
- Pump commanded on with normal pressure ....
- Pump with invalid data from sensor ....

Center Non-Normal Operation

Leak isolation logic protects key systems

If a leak is detected
- Two branches isolated from center system pressure
  - Nose gear actuation / steering
  - Leading edge slats
- One branch isolated and pressurized by C1 pump—**ALTN/RSV BRAKES**
- Remainder of center system
- Isolated and pressurized by C2 pump
- C1 and C2 demand pumps operate if required

Reservoir fluid quantity drop detected
- Flap/slat electronic Units prevent slats from operating in primary (hydraulic) mode
- Prevents slat retraction if loss due to catastrophic engine failure in critical phase of flight
- LE slats are reconnected to center system and operate in the normal mode If:
  - System volume recovers to normal for 5 seconds ***AND***
  - System detects both engines running more than 30 seconds

Reserve isolation valve remains closed for remainder of flight to isolate
- ♦ Main landing gear actuation & steering
- ♦ Flap & slat operation
- ♦ Center system flight controls

Nose gear isolation valve opens to restore pressure to nose gear actuation and steering if:
- ♦ Center system flight control pressure ***drops***
  - Leak is assumed to be in main landing gear, flaps or flight control lines
  - Nose gear isolation valve opens and C1 primary pump pressurizes nose gear branch
  - In this case, ALTN/RSV BRAKES branch remains pressurized

OR
- ♦ Center system flight control pressure ***is maintained***
  - Lowering gear handle opens **W** branch valve if both engines running and both primary engine-driven pumps are providing pressure
  - Ensures primary pump pressure is available for both left and right system flight controls and normal brakes before opening valve
  - Nose gear branch leak in this situation could cause loss of center system

OR
- ♦ Single engine operation
  - Above conditions would not be met
  - NG branch valve opens when AC slows below 60 knots
  - Allows steering when rudder effectiveness is lost

Ram Air Turbine

Priority Output

Provides hydraulic power to center system primary flight controls

Provides electrical power
- Power for standby system switches to Battery if RAT sheds electrical power generation
- See Electrical notes, *page 34* of this Study Guide for electrical functions of RAT.

Priority is hydraulic pressure over electrical power generation

Deployment

Deploys if: (HEED)
- **H***ydraulic—All 3 hydraulic system pressures are low* ***OR***
- **E***lectrical—Both AC transfer busses are unpowered* ***OR***
- **E***ngines both fail* ***AND*** *center hydraulic pressure low* ***OR***
- **D***eployed manually*

Manual deployment by pressing ram air turbine switch
- Hot battery or APU Battery bus must be powered
- Center hydraulic system does not need to be pressurized
- Deployment is by compressed spring
- RAT cannot be stowed in flight once deployed

Ram Air Turbine Switch

- Pressing the switch deploys the RAT if not already deployed
- **PRESS**—RAT is deployed ***and***
  - Center primary flight control hydraulic pressure is greater than 1500 psi
- **UNLKD**—RAT is ***not*** in the stowed position

Landing Gear

Key Components

Steering
- Nosewheel
- Aft pair on each main gear truck

Braking
- Main gear wheels only
- Normal brake pressure from right hydraulic system
- Alternate brakes on center hydraulic system
- Antiskid provided with all braking modes
- Autobrakes only available with normal brakes

Monitoring on GEAR synoptic
- Tire pressure readouts for each wheel
- Brake temperature monitors on each brake

Air/Ground sensing system

In-flight and ground operation of many systems changes depending on air or ground mode of airplane

Sensors located on each main landing gear beam

Landing Gear Operation

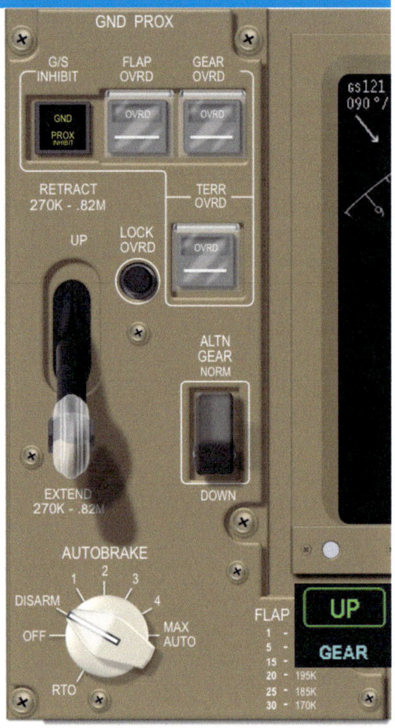

Gear Lever
- Automatic lever lock
  - secures handle down on ground
- In-flight lever lock is
  - automatically retracted through air-ground sensors
- Manual override possible
  - by pressing and holding the landing gear lever lock override switch

Retraction
- Sequence
  - Gear handle moved to up
  - Gear begins to retract
  - EICAS gear green **DOWN** indication changes to white crosshatch transition indication
  - Main gear doors open
  - Main gear wheels tilt to retract position
  - When retraction complete, indication changes to green **UP** indication
  - After 10 seconds, indication disappears
- Abnormal retraction
  - After normal time allowance for retraction, if retraction not complete—EICAS **GEAR DISAGREE** caution message displayed

  - Indication changes to expanded non-normal format
  - Affected gear displayed as in-transit down
  - Occurs if the gear is not unlocked from down position in normal time allotment
  - EICAS **GEAR DOOR** advisory message is displayed

Note: *WOW system can delay ground-air transition (as long as 17 seconds) if TO weight is light (<490K lbs) and 2 preceding landing weights were heavy (>525K lbs). This information removed from OM.*

Extension
- Sequence
  - Gear handle moved to down
  - Main gear doors open
  - EICAS gear in transition crosshatch appears

  - Gear free-falls without hydraulic pressure
  - Downlocks are powered to the locked position
  - Hydraulically operated gear doors close
  - Main gear trucks tilt to flight position
  - When gear are locked down, EICAS gear indication changes to **DOWN**

- **DOWN** indication means:
  - Side and drag brace on each main gear is in down and locked position ***and***
  - Nose gear drag brace is in down and locked position

Extension Abnormalities

- After normal time allowance for extension,
  - if any gear is not locked down—EICAS **GEAR DISAGREE** caution message displayed ***and***
  - Audible beeper warning sounds
- If gear does not lock down in expected time
  - Indication changes to expanded non-normal format
  - Affected gear displayed as in-transit down ***or***
  - **UP** indication if respective gear never unlocked
- If gear doors do not reclose in expected transit time, EICAS
  - **GEAR DOOR** message is displayed
- If only one brace on a main gear is locked
  - EICAS caution message **MAIN GEAR BRACE L** or **R** ***and*** audible beeper warning sounds
  - EICAS gear position indicator shows expanded non-normal display

Landing Gear Alternate Extension

Dedicated DC electrically powered hydraulic pump
- Uses center hydraulic system fluid
- Releases the uplocks on gear doors and landing gear
- Gear allowed to free-fall into position

Landing gear lever has no effect on system

EICAS
- Displays expanded gear position indication when using this system
- **GEAR DOOR** message is displayed because doors remain open

After alternate extension
- Landing gear may be retracted if the normal system is operating
- Select **DN** then **UP** on gear handle to retract

Nose Wheel and Main Gear Aft Axle Steering

Nose wheel steering tiller
- Allows nose wheel steering up to **70°**
- Powered by center system
- A pointer on the tiller shows the relative position

Rudder pedal steering—can steer nose wheel up to **7°**

Main Gear Aft Axle Steering
- Begins when nose wheel steering angle exceeds **13°**
- Reduces tire scrubbing
- Takeoff configuration warning message
  - **CONFIG GEAR STEERING**
  - Part of the takeoff configuration warning system
  - Indicates the main gear have not returned to a straight alignment when commanded to the center position when takeoff thrust is selected

Brake System

Main Gear Brake Features

Each wheel has a multiple carbon-disc brake set

Nose wheels have no brakes

Wear Indicator pins on each wheel must protrude beyond flush with parking brakes set

Normal Brakes

Right hydraulic system provides fluid to power brakes

Each brake pedal controls respective L and R wheel brakes

Alternate Brakes

Selection is automatic

If right system pressure is low
- Center / reserve system automatically supplies pressure
- Pressing a brake pedal sends hydraulic pressure through alternate antiskid valves to brakes

If center system quantity is low
- C1 primary pump is isolated from center hydraulic system
- Provides an alternate braking pressure source

Both right and center systems have low pressure
- **BRAKE SOURCE** light illuminates

EICAS advisory message **BRAKE SOURCE**
- Normal and alternate brake system pressures are low, ***AND***
- Reserve hydraulic source for alternate brake system is low

EICAS advisory message **RESERVE BRAKE / STRG**
The following may not be available:
- Alternate brakes from the reserve hydraulic source ***OR***
- Normal nose gear extension ***OR***
- Nose wheel steering

Brake Accumulator

Located in normal brake hydraulic system
Pressure is recharged by right system
Pressure is indicated on brake accumulator pressure indicator below captain's PFD
If right and **center / reserve hydraulic** systems are lost
Brake accumulator is available with no switch actions
Can provide several applications ***or***
Parking brake application

Anti-Skid

Works in all brake modes
Normal mode—Wheels independently controlled
Alternate brakes—Brakes controlled as follows:
Forward and center on each side of each truck paired
Aft truck wheels independently controlled

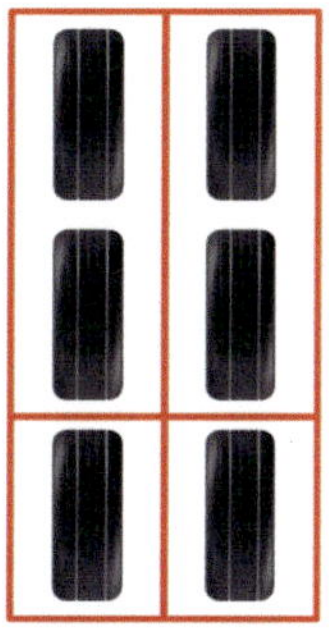

Control Information
Touchdown and hydroplaning protection uses inertial ground speed values
Locked wheel protection compares wheel speed to other wheels
EICAS ANTISKID message indicates one of the following:
Antiskid fault detected on brake system in use
Parking valve not fully open with parking brake released
System is completely inoperative

Autobrake System

Available only with normal brake system
Antiskid is provided with autobrakes
Selectable positions for landing

Armed by selecting prior to landing

Activated—braking application begins when
Both thrust levers are retracted to idle ***and***
The wheels have spun up

Selectable levels
AUTOBRAKE 1 through **4**
AUTOBRAKE MAX
AUTOBRAKE RTO (rejected takeoff)—takeoff position
Arming--Selecting prior to takeoff arms the system
Activating—RTO autobrake selects ***maximum braking*** pressure if:
Airplane is on the ground ***and***
Groundspeed is > 85 knots ***and***
Both thrust levers are retarded to idle
Maximum braking is obtained in this mode
If takeoff is rejected below 85 knots groundspeed, autobrakes do not activate

Autobrakes, if operative, must be armed prior to landing when any of the following exist:
- Runway:
  - Short: Runway length less than 7000 feet
  - Slippery: Braking conditions reported less than good
  - Contaminated: Runway contaminated with standing water, snow, slush, or ice
- Weather RVR less than 4000 or visibility less than 3/4 mile (Minimum setting—3)
- Approach speed increased above normal by procedure.

Autobrakes are recommended when landing with gusty winds or crosswinds
See AOM 9.1.1 Arrival Preparation Items, p. 9-3

Autobrake application is metered
Achieves programmed deceleration rates
As other deceleration means take effect, autobraking decreases to achieve targeted rates

For awareness only—according to instructor presentation:
Autobrake levels correspond to target deceleration rates in feet per second as follows:
Autobrake 1................................4 FPS deceleration
Autobrake 2................................5 FPS deceleration
Autobrake 3................................6 FPS deceleration
Autobrake 4................................7 FPS deceleration
Autobrake Max..........................11 FPS deceleration
Autobrake RTO ....Maximum possible deceleration

Note: *Any abort over 80 knots using maximum braking (RTO) requires maintenance inspection prior to subsequent takeoff ....QRH LDG GEAR 14.6, BRAKE TEMP*

Use in normal conditions—"For normal landing conditions, autobrakes 2 or 3 will optimize brake wear, passenger comfort and stopping performance."........AOM 9.1.1 Preparation Items, p. 9-3

Taxi brake release
Feature active when taxi speed less than 45 knots
Antiskid releases one axle pair of main gear brakes during each brake application
Axle pair is sequentially changed on each application
Extends brake life and reduces brake sensitivity
Feature is NOT active for any of the following:
Any brake system except normal is in use
"Heavy brake application" occurs (not defined)
Landing rollout
Rejected takeoff
Parking brake applied

AUTOBRAKE Advisory message—indicates Autobrake disarmed or inoperative

Braking during landing
Five levels of deceleration available
1 through **4**, and **MAX**—see above
On dry runway, full pedal braking applies more deceleration than any autobrake *landing* mode
AUTOBRAKE MAX
Results in **AUTOBRAKE LEVEL 4** until pitch angle <1°
Then **MAX AUTO** level braking commences
Deceleration level can be changed without disarming system by rotating selector
To maintain selected deceleration rate, brake pressure is varied as other deceleration forces are applied (reverse thrust, spoilers, etc.)

Deactivation—Autobraking stops when any of the following occur:
- Full stop is attained (autobrake mode remains engaged)
- Pedal brakes are applied
- Either thrust lever is advanced after landing
- Speedbrake handle is moved to the **DOWN** detent after speedbrakes have deployed on the ground
- **DISARM** or **OFF** is selected on the autobrake selector
- Autobrake fault
- Normal antiskid system fault
- Loss of inertial data from ADIRU
- Autobrakes are applied after loss of normal brake hydraulic pressure

When disarmed after landing
Selector moves to **DISARM** position
Power is removed from the autobrake system
When disarmed after takeoff
Selector remains in **RTO**
Automatically moves to **OFF** after takeoff

Parking Brake

Can be set with either normal or alternate system pressurized
If neither normal nor alternate are available, parking brake pressure is maintained by brake accumulator
When set, EICAS memo message **PARKING BRAKE SET**
If set when either engine is advanced to takeoff thrust
Takeoff configuration warning message **CONFIG PARKING BRAKE**
Aural alert sounds

Brake Temperature Indication

Digital values rate heat on each brake from 0 to 9.9 scale
Normal temperature range is 0.0-4.9
Empty white box for temperature range 0-2.9
Filled white box
Shown when any wheel's temperature is in the high end (3.0 - 4.9) of the normal range
Highest temperature wheel on each truck is shown with filled box
Filled amber box
"High range" value, 5.0 - 9.9
EICAS advisory message **BRAKE TEMP** for values of 5.0 and higher

Automatic Relay Summary for Takeoff & Landing

	Takeoff	Aborted Takeoff	Landing	Go-Around
Auto Brakes Arming & Activation	Armed in **RTO** position	Activated when: ♦ Airplane is on the ground *and* ♦ Groundspeed is > 85 knots *and* ♦ Both thrust levers are retarded to idle	**Brakes applied when:** ♦ Both thrust levers are retracted to idle *and* ♦ The wheels have spun up	Brakes released when either thrust lever is advanced (*See below*)
Disarming & Deactivation		♦Full stop is attained (autobrake mode remains engaged) ♦Pedal brakes are applied ♦Either thrust lever is advanced after landing ♦Speedbrake handle is moved to the **DOWN** detent after speedbrakes have deployed on the ground ♦**DISARM** or **OFF** is selected on the autobrake selector ♦Autobrake fault ♦Normal antiskid system fault ♦Loss of inertial data from ADIRU ♦Autobrakes are applied after loss of normal brake hydraulic pressure		
Auto Spoiler	No arming	Either thrust reverser selected deploys ground spoilers	Armed when handle placed in ARMED position inflight Activated when: ♦ Gear on ground (not tilted) *AND* ♦ Thrust Levers both IDLE *OR* ♦ Either thrust lever moved to reverse idle detent	Retracted when either thrust lever is advanced to **TAKEOFF** range Handle moves to down/ unarmed position

EICAS Display—Brakes & Tire Pressures

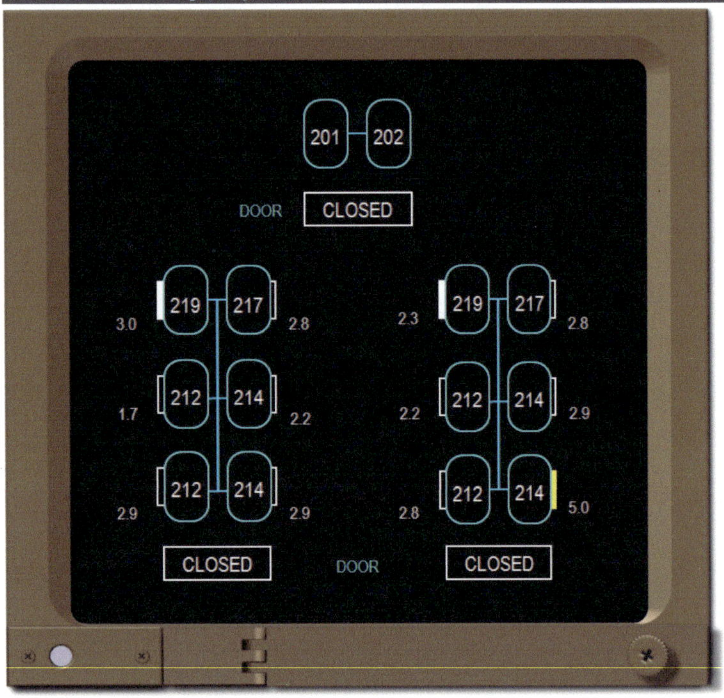

Tire Pressure Indication

Individual tire pressures displayed on gear synoptic
Allowable range is 0-400 psi
EICAS advisory message TIRE PRESS
- Indicates any tire is above or below normal range
- Also indicates excessive difference between two tires on same axle

300ER Tail Skid

Extends / retracts when landing gear handle is lowered / raised
EICAS Advisory TAIL SKID displayed when not in the correct position

Fault Indications

- BRAKE—brake deactivation on associated wheel
- ASKID—antiskid fault on associated wheel
- Both appear on gear synoptic next to appropriate wheel

Warning Systems

Engine Indication and Crew Alerting System (EICAS)

General Features

Consolidates engine and airplane system indications
Primary display for system alerts and indications
Normally displayed on upper center display
Primary engine indications are also on upper center display

Master Warning and Caution Reset Switches & Lights

Press—Extinguishes Master Warning and/or Master Caution light(s)
- Silences aural tones associated with these EICAS warning messages:
  - ♦ **CABIN ALTITUDE**
  - ♦ **FIRE**
  - ♦ **OVERSPEED** (some A/C)
  - ♦ **STABILIZER**
  - ♦ **CONFIG GEAR**, if displayed because landing gear not down and locked, any thrust lever at idle, and radio altitude < 800 feet

Relights for each new Warning / Caution as they are initiated

EICAS Event Record Switch

Records up to five EICAS events into memory
- When button pressed, records
  - Currently displayed engine indications *AND*
  - Additional EICAS maintenance information
- Without button pressed, if a system parameter is exceeded, automatically records out of limit parameters and related conditions

Located on forward aisle stand

EICAS Messages

Alert messages—crew alerting for non-normal conditions
- From engine start to shutdown—primary means of crew alert
- EICAS checklist steps are included in QRH documentation
- Rectangle ◻ appears left of the message if:
  - An alert has a checklist associated *OR*
  - No checklist, but notes or information of which crew needs to be aware

Alert Message Priority Levels
- Warning .... ◻ RED

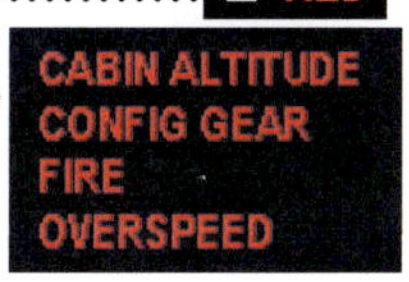

  - Cannot be cancelled
  - Accompanied by ***aural bell*** for fire warnings
  - Accompanied by ***siren*** for warnings of:
    - ♦Autopilot disconnect
    - ♦Cabin altitude
    - ♦Overspeed (some airplanes)
    - ♦Takeoff configuration warnings
  - Voice warnings for some other conditions
  - Preceded by a box ◻ indicating checklist steps exist if not completed or superseded by another procedure
- Caution Message .... AMBER
  - New caution accompanied by ***beeper*** except for **ENGINE SHUTDOWN L** or **R**

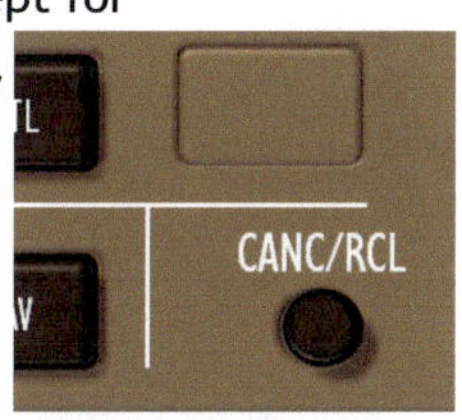

  - May be cancelled by pressing **CANC/RCL** switch
  - May also be recalled by pressing the switch again
  - Note: *References to the* **CANC/RCL** *button in this section are for the display select panel button. Do not confuse with the* **CANC** *button on the ACARS message management button set.*

- Advisory Message .... AMBER
  - Next highest alert message
  - Indented one space from Warning and Caution messages
  - May be cancelled or recalled by pressing **CANC/RCL**
- Communication messages .... • WHITE
  - Preceded by white dot • and indented one space
  - Cannot be cancelled or recalled by pressing **CANC/RCL**
  - Three priority levels—all are white in color
    - High—***Chime*** associated with alert
    - Medium—***High-low chime*** for new message
    - Low—No chime associated
  - All communication messages require crew action
    - Usually involves selecting Comm to display messages
    - Accomplishing action required makes message go away
    - Message cannot be cancelled other than by required action
  - If several COMM messages, most recent at ***top*** of list
- Memo Messages .... WHITE
  - Status of selected flight-crew activated normal conditions
  - Appear at bottom of last page of EICAS alert messages
  - Pressing **CANC/RCL** button when last page of EICAS alert messages is displayed—ensures all current memos have been displayed
  - If several memo messages, most recent is at ***bottom*** of list
- Status messages .... CYAN
  - A/C equipment faults
  - Faults so displayed are those affecting airplane dispatch
  - Displayed on MFD **STATUS** page
  - **STATUS** cue displays on primary EICAS display when new message appears
  - If several status messages, most recent is at ***top*** of list
  - **STAT** button
    - Displays single and multiple pages of status messages
    - Functions for status messages similarly to **CANC/RCL** button for EICAS messages

Display and Manipulation of Messages

Grouped by priority
- **WARNING** messages (red) at top of list
- **CAUTION** messages (amber) below last warning message
- **ADVISORY** messages Indented one character to further demonstrate lower priority
- Within groupings, most recently appearing message is at top of list
- If a condition indicated by a message no longer exists
  - Message goes away ***AND***
  - Items below move up one line to fill in

If more messages exist than can be displayed
- Multiple pages are created
- Pages are numbered, with numbers at bottom of list
- If multiple pages of Caution and Advisory messages
  - Pages can be advanced with **CANC/RCL** button
- If more Warning conditions exist than can be displayed
  - No page number shown
  - Pages ***cannot*** be advanced with **CANC/RCL** button
  - No other pages can be displayed
- When at last page of caution and advisory messages
  - Pressing **CANC/RCL** button makes last of messages disappear and page go blank
  - Pressing again brings back first page of messages

Airspeed Alerts

Stall warning

Stick shakers used to alert crew to condition
Independent shakers on each control column

Airspeed low

Occurs when speed below min maneuvering
EICAS Caution **AIRSPEED LOW**
Amber box appears around current speed

Takeoff V1 Airspeed

VEE ONE sounds on takeoff when
- Speed reaches V_1

Overspeed Warning

Occurs when speed is greater than V_{MO}/M_{MO}
EICAS message **OVERSPEED** is displayed
Message remains until condition goes away

Tail Strike Detection System

Detects ground contact that could damage airplane pressure hull
Sensors
- Two proximity sensors
- Two-inch blade antenna
- Installed on aft body of airplane

If detected, EICAS caution message, **TAIL STRIKE**
Tail strike protection through flight controls
Note: *For other flight envelope protection details see FLT-C section, page 41 of this book*

Takeoff & Landing Configuration Warnings

Takeoff Configuration Warning

EICAS Warning message **CONFIG** is displayed if the following are all met:
- ♦ **Ground**—Airplane is on the ground
- ♦ **Run**—Fuel control switches are in Run
- ♦ **Thrust**—Either engine thrust is in takeoff range
- ♦ **Speed**—Airspeed is less than V1

AND Any one of the following configurations exist:
- ♦ **Doors** (any) not closed, latched and locked
- ♦ **Flaps** not in takeoff position
- ♦ **Main gear steering** not locked
- ♦ **Parking brake** set
- ♦ **Rudder trim** not centered
- ♦ **Speedbrake** handle not down
- ♦ **Stabilizer trim** not in green band

Inhibited at V_1
When message displays
- Siren sounds and master **WARNING** lights comes on
- Pressing **WARNING / CAUTION RESET** switch
  - Resets the master warning lights
  - Does not silence siren
- Siren remains on until
  - Condition no longer exists *OR*
  - Thrust Levers are retarded below takeoff range
- CONFIG message remains on 10 seconds after thrust levers retarded below takeoff range
  - Aids in determining cause of siren

Landing Configuration Warning

Alerts the crew if the landing gear is not extended for landing
EICAS Warning message **CONFIG GEAR** is displayed if the following are all met:
- ♦ Airplane is in flight
- ♦ Any landing gear is not down and locked

AND Either one of the following configurations exist:
- ♦ Radio altitude is less than 800 feet and either thrust lever is closed
- ♦ Flap handle is in landing position

If message is displayed because thrust lever is retarded at low altitude
- Pressing either master **WARNING** / **CAUTION** reset switch
  - Silences siren
  - Extinguishes master **WARNING** lights
- Message remains displayed until
  - Thrust Levers advanced *OR*
  - Landing gear is down and locked

If message is displayed because flap handle is in landing position without gear down and locked
- Siren cannot be silenced until
  - Gear down and locked *OR*
  - **GND PROX GEAR OVRD** switch is pressed

Message remains displayed as long as condition exists

Non-Normal Operation of Configuration Warning System

If system fails, EICAS advisory **CONFIG WARN SYS** message displayed
If TO and landing configuration system fails, **CONFIG** messages may or may not be displayed
If messages appear along with the **CONFIG WARN SYS** messages, the **CONFIG** messages may not be correct

MCP Selected Altitude Alerts

Purpose—Notify crew when approaching or departing altitude selected in MCP altitude window
Approaching MCP-selected altitude
- 900' Before reaching selected altitude—white box appears around selected altitude and current altitude on PFD
- 200' Before selected altitude—white boxes go away

Departing MCP selected altitude by 200 feet
- EICAS alert message **ALTITUDE ALERT**
- Amber box appears around current altitude
- Message and amber box disappear when:
- Altitude returns to within 200 feet of selected altitude *OR*
- New MCP altitude is selected *OR*
- Altitude continues to increase to more than 900 feet from MCP altitude selection

Altitude alert inhibits occur when
- Glideslope captured *OR*
- Landing flaps selected and landing gear down and locked

Traffic Alert & Collision Avoidance System (TCAS)

Controlled through Transponder Panel

TCAS Operating Policy

Both CA & FO must be qualified to operate
Visual contact made with traffic--See & Avoid
TCAS RA Issued—
- MUST comply with warning, even if traffic in sight
- Assures mutual collision avoidance actions occur, as coordinated through TCAS

Only vertical avoidance maneuvers are permitted in response to an RA
- If in a turn , continue the turn
- If "*INCREASE CLIMB*" RA occurs in a turn, level wings and climb at faster rate

Pilot Flying makes smooth, controlled responses
Assumed reaction time:
- 5 Seconds from RA command issuance
- 2.5 Seconds from reversal command for existing RA

No Enforcement action is to be taken against flight crews complying with TCAS who deviate from ATC clearance

Altitude Band Selection

ABOVE (ABV)—Used in Climb
- Traffic -2700 to +7000 from present altitude

Normal (N)—Used in cruise
- Traffic ±2700' from present altitude

Below (BLW)—Used in Descent
- Traffic +2700 to -7000' from present altitude

Traffic Symbols and Warnings

Other Traffic ....**Open White Diamond**
- **More than 6 miles away** *OR*
- Over 1200' altitude differential

Proximate Traffic ....**Filled White Diamond**
- **Within 6 Miles** *AND*
- Within ± 1200' altitude

Traffic Advisory (TA) Intruder ......**Filled Amber Circle**
- Predicted path of another airplane which will enter TCAS conflict airspace within 20-48 seconds
  - *Aural Warning* ....*TRAFFIC TRAFFIC*"

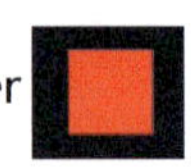

Resolution Advisory ....**Filled Red Square**
- Predicted path of another airplane which will enter TCAS conflict airspace within 15-35 seconds
- Threat is 25 seconds from closest point of approach
- Vertical avoidance maneuver displayed on IVSI
- If range farther than current ND scale, **OFFSCALE** displays on ND
  - ASM 10.40.21; 15.10.7 through 15.10.9, and 15.10.19
  - Aural Warning—None listed ....See AOM 15.20.6-to 15.20.8

Required flight crew response
- Within 5 seconds of initial RA
- Applying G-forces of ±.25 G
- This is more than AP can supply, so response must be hand-flown (*no change from previous TCAS policy*)

Ground Proximity Warning System (GPWS)

GPWS Functions

Alerts crew of predicted ground proximity problems in any of six modes:

Mode 1—Excessive descent rate below 2450' AGL

Mode 2—Excessive terrain closure rate below 2450' AGL
- See also modes 2A and 2B

Mode 3—Altitude loss after takeoff or go-around between 30 and 1330' AGL

Mode 4—Unsafe terrain clearance while not in landing configuration (either mode 4a or 4b)

Mode 5—Below glide slope more than 1.3 dots while below 1000'AGL

Windshear Mode—Significant windshear < 1500' AGL

Inhibit Switches

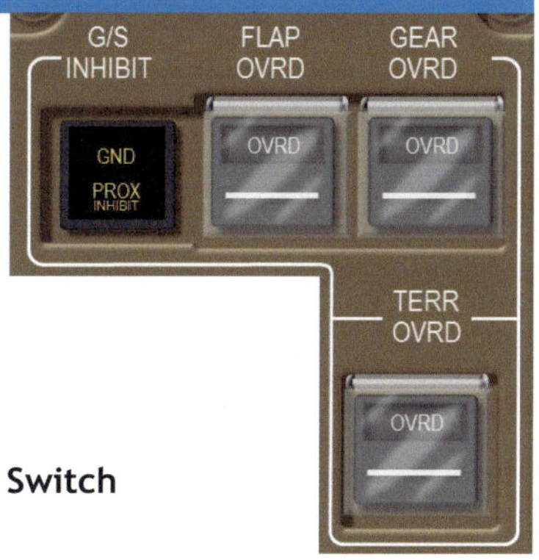

G/S Inhibit Switch
- Pressed—Inhibits ***BELOW GLIDE SLOPE*** alerts when pressed below 1000' RA
- **GND PROX**—Indicates GPWS alert occurring

Ground Proximity Gear Configuration Override Switch
- Pressed—(**OVRD**—*Amber*)
- Inhibits alerts for
  - ***TOO LOW GEAR***
  - ***CONFIG GEAR***

Ground Proximity Flap Override Switch
- Pressed—(**OVRD**—*Amber*)
- Inhibits ***TOO LOW FLAPS*** **alert**

Ground Proximity Terrain Override Switch
- Pressed—(**OVRD**—*Amber*)
- Inhibits look-ahead terrain alerts and display

Enhanced GPWS (EGPWS)

Major features
- Worldwide database with all runways over 3500' long
- (Not all of these can be displayed in map mode)
- Terrain data from worldwide database can be shown
- Warning capability if airplane flight path predicted to intersect terrain within specified criteria

Display
- Variable density dot patterns
- Dot density indicates height in respective color zone
- **ND TERR** message annunciated when active

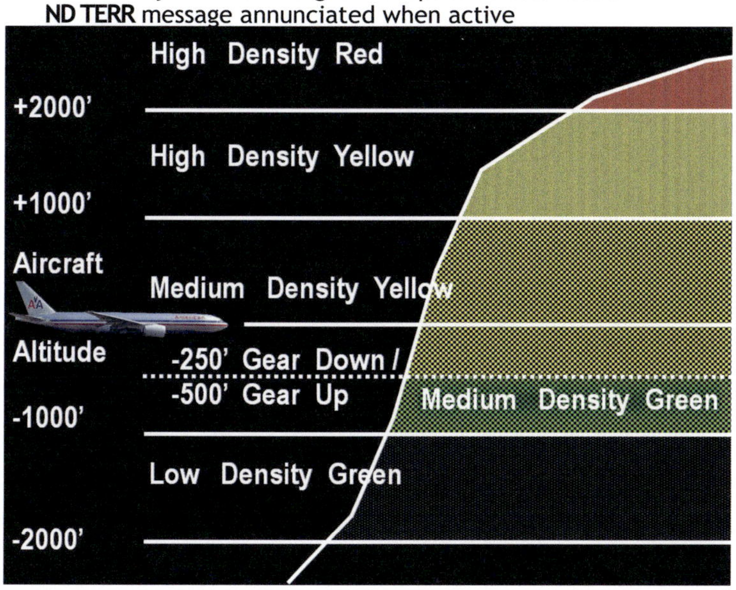

Alerts
- Caution Alert
  - Criteria—40-60 seconds from predicted terrain conflict
  - Conflict criteria
    - Within about 1/8 mile either side aircraft
    - The initial distance of 1/8 mile (1/4 mile wide) expands ±3 from both sides of aircraft as the path is projected forward
- Warning Alert
  - Criteria—20-30 seconds from predicted terrain conflict
- Approach & departure
  - Criteria modified to avoid nuisance alerts

Weather Radar

General

Primarily turbulence avoidance tool

Echoes are from rainfall and water in atmosphere; turbulence is implied

Turbulence mode
- Turbulence directly detected by measuring rainfall velocity
- Applies to storms within selected range of 40NM or less

Windshear detection
- Events detected within 3 NM in front of aircraft
- Referred to as predictive windshear

Radar antenna is stabilized with horizon by ADIRU inputs

Weather Radar System

Weather returns displayed in all modes except:
- Plan
- VOR Centered
- Approach Centered

ON/OFF Switch is EFIS WXR map switch

Variable gain settings available in **WX**, **WX+T** or **MAP** modes

Testing
- Weather radar and predictive windshear system may be tested simultaneously
- Select **WXR** on EFIS panel and then press **TEST** switch
- Activates test of both systems

Scan Pattern
- Normal modes— ±90°
- Predictive windshear mode below 2300'AGL— ±60°
- Tilt—adjustable ±15° from horizon
- May be selected differently on L and R sides
- If tilt angles are different, an audible "bump" may be heard as antenna angle adjusts
- Antenna scan is stabilized only to 40°, so if pitch plus roll exceeds this amount, ground return on one side of the display

Display update
- Every sweep if displayed on only one ND
- Alternate sweeps if displayed on both NDs

PAC Alert
- Annunciation identifies areas of severe attenuation
- Indicates signal is attenuated below min discernible signal level
- Yellow arc paints on outer edge
- Requirements:
  - Gain control in calibrated position (12 0'clock)
  - Attenuating position is 80NM or less from A/C
- Areas between painted return and PAC alert bar must be avoided

Ground Clutter Suppression (GCS)
- Removes approximately 85% of ground clutter targets
- Also removes some stationary valid weather returns
- Recommended use is temporary for identification of ground targets
- Works best with shallow tilt angles of 0-5° tilt down

GPWS Predictive Windshear System (PWS)

System Activation
- Enabled during takeoff, approach and landing
- Provides immediate alert for the following:
  - Excessive downdraft or tailwind
  - Excessive windshear detected ahead of airplane

Master WARNING light accompanies warning
- Pressing a master **WARNING** light extinguishes both lights
- Windshear alerting is *not* inhibited by pressing master **WARNING**

System Operation

- Uses weather radar to detect disturbed air in front of airplane
- Weather radar automatically scans in PWS mode when:
  - On ground with thrust levers in takeoff range ***OR***
  - In-flight below 2300' radio altitude
- Alerts enabled approximately 12 seconds after scan begins
- Prior to takeoff—Alerts can be enabled by pressing WXR switch on EFIS panel
- During approach—Alerts enabled below 1200' radio altitude
- Radar sweep reduced to ±25° when PWS active
- Radar returns only display (even though PWS is active) if WXR switch is selected

PWS Alert Area

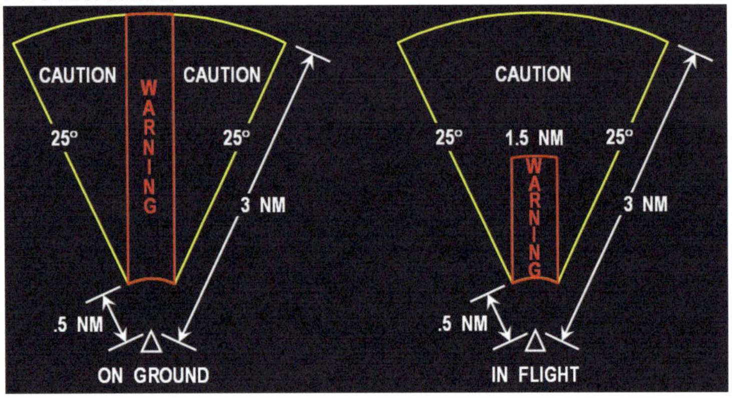

Alert Inhibits

Inhibit Philosophy—Inhibited when:

Operationally unnecessary or inappropriate

- Many alerts indicate abnormal situations only if system is on
- Example—No warning for low oil pressure before engine start

Condition normal for system operation

- Time delays for in transit indications
- Example is valves in transit—no warning to indicate not in commanded position until after time delay

Prevents distraction during takeoff and landing

Voice Annunciation Inhibits

Warnings inhibit related cautions

Voice alerts are prioritized and most important priority messages override lower priorities

Takeoff and Landing Inhibits

Takeoff

- All but most important alerts dealing with engine failure and PWS inhibited from T/O thrust set to V_1
- Inhibits are removed at 400' through 1100' in stages

See ASM 15.20.40, p.15-20-34 through 15-20-39

Landing

- Inhibits progressively increase below 1500'RA on approach
- Inhibits are removed at 100, 80 and 75 knots

See ASM 15.20.40, p.15-20-40

Acronyms

AIMS ....airplane information management system
EICAS ....engine indication and crew alerting system
GCS ....ground clutter suppression
GPWS ....ground proximity warning system
PWS.......................................predictive windshear system
TCAS.....................traffic alert and collision avoidance system

Titles Currently Available

By Rick Townsend

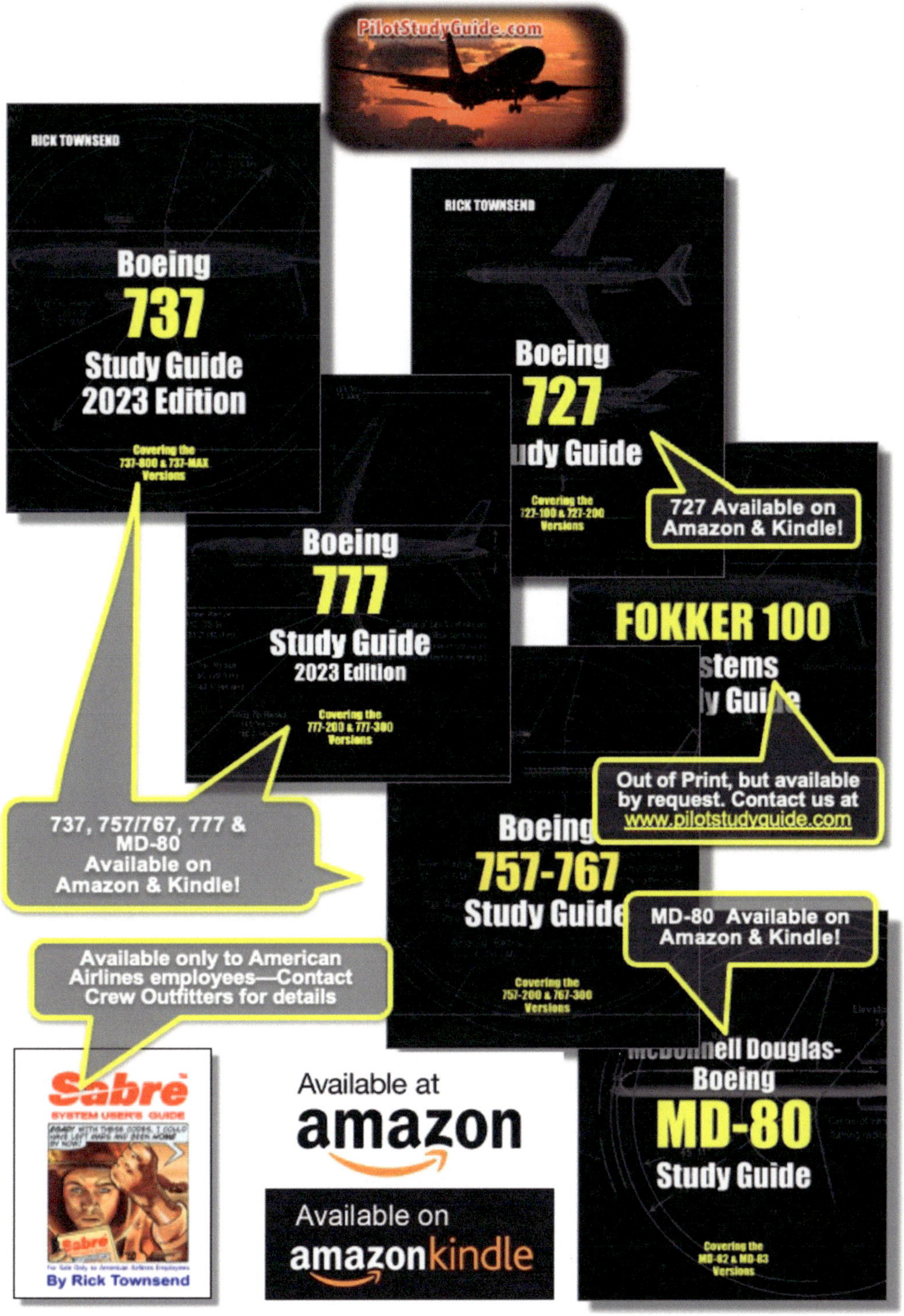

For more information on these books, see our web site for Pilot Study Guides

To purchase any of these titles, go to Crew Outfitters locations at the Flight Academy Gift Shop, or at DFW Airport Terminal C Operations
To purchase online, go to *www.crewoutfitters.com*

To order by phone, call **800-874-1204**

This book is available through Amazon in both Kindle and in print format!

Giving Back

A portion of the proceeds from every book help support some worthy charities. Here are two of the charities your book purchase supports:

Kenya Kids Can!

Feeding and teaching world changers of tomorrow

Kenya Kids Can is a non-profit working in the Rift Valley of Kenya. They feed students and construct solar powered computer centers to improve education and bring hope to needy students. Please browse the website to learn about what can be done to truly change lives.

www.kenyakidscan.org

Gary Sinise Foundation — Snowball Express

GARY SINISE FOUNDATION
SNOWBALL EXPRESS
PRESENTING SPONSOR American Airlines

Serving the *children* of our fallen military heroes
Snowball Express Could Use Your Help!

Snowball Express honors America's fallen military service members who have made the ultimate sacrifice since 9/11 by

- **Humbly serving the families they left behind**
- **Championing their children's future success by creating opportunities for joy, friendship, and communal healing**
- **Connecting these families to one another.**

Since 2006, the mission of Snowball Express has been a simple, yet profoundly important one: Providing hope and new happy memories to the children of military fallen heroes who have died while on active duty since 9/11. In December each year we bring children together from all over the world for a four-day experience filled with fun activities, like sporting events, dances, amusement parks and more.

Nationally, Snowball Express provides comprehensive support programs for fallen families that are focused on transition and connections to community resources, healing and wellness, peer engagement, education and personal/professional development programs.

https://www.garysinisefoundation.org/snowball-express

Made in the USA
Monee, IL
12 September 2026

62974522R00040